SpringerBriefs in Applied Sciences and Technology

SpringerBriefs present concise summaries of cutting-edge research and practical applications across a wide spectrum of fields. Featuring compact volumes of 50 to 125 pages, the series covers a range of content from professional to academic.

Typical publications can be:

- A timely report of state-of-the art methods
- An introduction to or a manual for the application of mathematical or computer techniques
- A bridge between new research results, as published in journal articles
- A snapshot of a hot or emerging topic
- An in-depth case study
- A presentation of core concepts that students must understand in order to make independent contributions

SpringerBriefs are characterized by fast, global electronic dissemination, standard publishing contracts, standardized manuscript preparation and formatting guidelines, and expedited production schedules.

On the one hand, **SpringerBriefs in Applied Sciences and Technology** are devoted to the publication of fundamentals and applications within the different classical engineering disciplines as well as in interdisciplinary fields that recently emerged between these areas. On the other hand, as the boundary separating fundamental research and applied technology is more and more dissolving, this series is particularly open to trans-disciplinary topics between fundamental science and engineering.

Indexed by EI-Compendex, SCOPUS and Springerlink.

Santosh Jagtap

Design and Engineering for Low Resource Settings

A Practical Guide

 Springer

Santosh Jagtap (ID)
Department of Design
Indian Institute of Technology
Guwahati, Assam, India

ISSN 2191-530X   ISSN 2191-5318   (electronic)
SpringerBriefs in Applied Sciences and Technology
ISBN 978-3-031-66155-6   ISBN 978-3-031-66156-3   (eBook)
https://doi.org/10.1007/978-3-031-66156-3

This Springer imprint is published by the registered company Springer Nature Switzerland AG
The registered company address is: Gewerbestrasse 11, 6330 Cham, Switzerland

If disposing of this product, please recycle the paper.

Preface

In the pursuit of development and progress, it is important to ensure that no one is left behind. Specifically, in resource-constrained settings, where challenges are abundant and conventional solutions may not be suitable, innovative approaches are needed. This book explores such approaches, focusing on a practical and integrated methodology for designing, developing, and implementing solutions for the betterment of people living in low resource settings.

Chapter 1 presents the background by examining various design and engineering approaches such as appropriate technology, humanitarian engineering, design for development, base of the pyramid, frugal innovation, and social innovation. It emphasizes the necessity of holistic solutions for addressing infrastructure weaknesses, limited resources, and low literacy levels and proposes a comprehensive and integrated methodology to guide the process of designing, developing, and implementing solutions aimed at the betterment of resource-constrained societies.

Chapter 2 highlights the critical role of local context in designing integrated solutions for low resource settings and presents guidelines for understanding, adapting to, and building on local conditions and circumstances. By providing illustrative examples, it demonstrates how a deep understanding of the socio-cultural context and adaptation of solutions can lead to more impactful solutions.

Chapter 3 examines the role and necessity of codesign with individuals living in resource-constrained societies and highlights its advantages for the design process as well as for human and social development of these societies. The chapter offers guidelines for effective codesign practices. These guidelines can support inclusive and impactful cocreation with underserved communities.

Taking into account the multifaceted needs of resource-constrained communities, Chap. 4 offers guidelines for designing training, awareness, and income generation programmes. By tailoring solutions to local contexts and circumstances, sustainable and appropriate solutions can be developed to support development of communities.

The final chapter (Chap. 5) examines the critical role of considering the entire lifecycle of solutions and fostering collaboration across sectors in the process of

designing and implementing solutions. By providing examples and recommendations, it highlights the significance of context-specific project management and
partnership-driven approaches in designing for sustainable and long-lasting impact.

In navigating these chapters, readers will embark on a journey through innovative methodologies and practical insights aimed at empowering communities and
advancing social and human development in resource-constrained settings.

Collectively, the book presents insights and best practices, which can inspire action
and foster positive change in the lives of those most in need. It is hoped that the ideas
and methodologies discussed in the book will contribute to a more equitable and
inclusive future for all.

I would like to thank Anuska Sahu for contributing to the creation of some visual
illustrations and Akanksha Dixit for some figures used in the book.

Guwahati, India   Santosh Jagtap

Contents

Chapter 1
Design and Engineering for Low Resource Settings: An Integrated Methodology

Abstract Design plays a crucial role in satisfying unmet or underserved needs of resource-constrained individuals and societies. This chapter presents an overview of various design and engineering approaches, such as appropriate technology, design for the real world, base of the pyramid, frugal innovation, and social innovation, aimed at designing and developing appropriate solutions adapted to specific conditions and deprivations encountered by resource-constrained population. It highlights the critical need of designing and developing integrated solutions that address multi-dimensional challenges such as weak infrastructure, limited resources, and poor literacy levels, while taking into account strengths of these resource-constrained societies. The chapter then outlines ten guidelines for devising such integrated solutions and presents an integrated methodology by mapping these guidelines over a design process.

1.1 A Design Process Model

Design, with its essential idea of changing an existing situation into a desired situation, is indispensable to create solutions that satisfy needs and requirements in a specific context (Simon 1996; Jagtap 2019b). Further, it plays an important role in creating solutions for a diverse range of population, including those living in resource-limited societies. A number of design activities need to be undertaken in order to create such solutions. Models of design processes provide a structured and systematic support for designing solutions by organising such design activities in some phases. There are several advantages of a systematic design process as listed below (Pahl et al. 1996; Jagtap et al. 2014b).

- It allows to explore and define aims and objectives of the project, while ensuring that all relevant stakeholders have a commonly shared understanding of these aims. Furthermore, it supports teams in developing and implementing schedules and targets that are grounded in reality.

© The Author(s), under exclusive license to Springer Nature Switzerland AG 2024
S. Jagtap, *Design and Engineering for Low Resource Settings*,
SpringerBriefs in Applied Sciences and Technology,
https://doi.org/10.1007/978-3-031-66156-3_1

- It helps design teams in identifying possible factors that can have negative conse- quences on the project success, and thereby, in devising ways to eliminate such factors or to alleviate their potential negative impact.
- It typically enhances communication among different stakeholders, supporting them to effectively contribute towards design tasks. Additionally, employment of a systematic design process supports team members to document information, including various decisions made and rationale behind these decisions. This can further enhance communication and transparency.
- Since various activities are organised in a structured design process, it helps to effectively and efficiently use time, effort, and other resources for designing, developing, and implementing a solution.
- A systematic design process can potentially support consistent use of procedures and activities.

Several models of design processes exist (e.g., Pahl et al. 1996; Ulrich et al. 2020; Cross 2023). There are some differences between such models regarding the termi- nology used to describe design activities, labels employed to describe phases for organising these activities, details of these activities and phases, and their graph- ical representation. Despite such differences, they share several commonalities. In general, a model of a design process consists of the following three phases: (1) understanding context and needs, (2) concept exploration and detailing, and (3) implementation. These phases are elaborated in the paragraphs that follow.

Phase-1: Understanding context and needs

This initial phase has a profound impact on the later phases and the resulting solution. This phase involves the following activities (see Kumar 2012):

- Gaining deeper understanding and insights into ongoing and emerging trends in the world, including those in technology, culture, environmental issues, economy, legal aspects, etc.
- Gathering information on the recent happenings, advancements, and current news.
- Gaining insights into the target context in which the eventually developed solutions might be implemented.
- Gathering crucial insights into the lives of people—target users as well as other relevant stakeholders—and the ways in which they interact with various aspects in their daily routines.
- Structuring the insights and understanding gained in the above activities to reveal key patterns and to identify crucial opportunities for designing appropriate solutions.

Phase-2: Concept exploration and detailing

Similar to the initial phase, this phase also consists of a number of activities. These are as follows (see Kumar 2012):

- A key activity in this phase is to generate concepts. The generated concepts need to be appropriate to the target context, and therefore, ought to be based on the results of the earlier phase.
- Concepts for fulfilling a diverse range of functions are generally explored. These functions can be about marketing, distribution, various life cycle issues of the planned solution, and other pertinent requirements in the target context.
- The generated concepts are evaluated taking into account relevant criteria, including their suitability for the target context.
- The most valuable and appropriate concepts are integrated to develop holistic solutions. The development of these holistic solutions can require iterative prototyping as well as testing in the actual setting.
- This phase also includes detail design of the generated concepts and solutions. This includes specifying, for example, dimensions and materials in the case of products. The detail design can also consist of specifying details of various aspects of solutions, which can be about marketing, distribution, and relevant aspects related to the life cycle of the solutions.

Phase-3: Implementation

This phase consists of the following activities (see Kumar 2012):

- In the final phase, the solutions generated in the previous phase are developed further for their realisation and implementation in the target context. This necessitates creation of roadmaps for their realisation and implementation.
- The creation of the above roadmaps can be a collaborative activity involving a number of stakeholders. The roadmaps are shared with the relevant stakeholders, providing them clear and specific information on the steps required to realise and implement the solution.
- In this phase, it is also necessary to gather requisite resources, formulate budgets and schedules, and recruit teams.

The design process model consisting of the above phases is shown in Fig. 1.1.

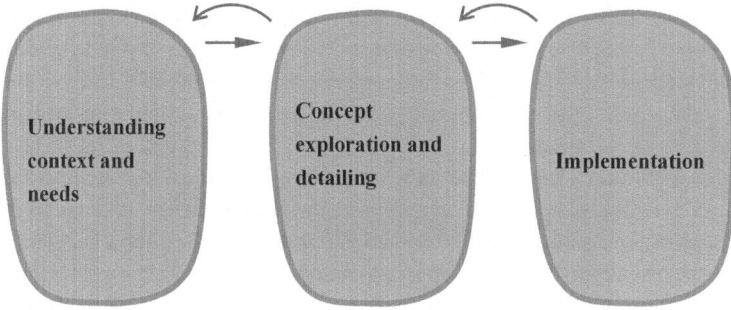

Fig. 1.1 Design process model

The term 'process' might suggest steps than need to be followed sequentially. However, this can be misrepresentation of the process since design and development projects are typically nonlinear (Kumar 2012). For instance, an idea generated during a brainstorming session might be the start of a project. The project can then include activities from the initial phase to support and refine the idea, followed by other activities.

In addition to the 'nonlinearity', the process is also 'iterative' in nature (e.g., Kumar 2012). This necessitates several cycles not just through the process but also through a set of some activities in a specific phase of the process. For example, a project might begin with gaining insights into the daily lives of the people in a specific context, followed by identification of opportunities to design a solution, and then concept generation. The project can then proceed to initial phase of the process to gain deeper understanding of the ways in which the people in the context use some specific products. This can then be followed by further concept exploration, user feedback, prototyping, testing in actual setting, and so on. The number of iterations in a design and development project can depend on available resources as well as the nature and scope of the project (e.g., Kumar 2012). While some projects might require a large number of iterations, other may need just a few iterations.

1.2   Low Resource Settings: Role of Design

People living in low resource settings in developing countries face multidimensional issues. These issues relate to poor health, unfreedom, high rates of mortality, and low and intermittent income (e.g., Jagtap 2019a, b; Rivera-Santos and Rufín 2010). These people can suffer from poverty and marginalisation and encounter a wide array of challenges such as limited income-generating prospects, restricted access to financial facilities, feelings of social isolation, mental stress resulting from a lack of literacy, exhaustion from strenuous physical labour, and inadequate or lack of access to fundamental amenities like healthcare, education, clean drinking water, infrastructure, energy, and sanitation (e.g., Jagtap et al. 2014a; Aranda-Jan et al. 2016; Prahalad 2004; Jagtap 2019a, b). Before proceeding further, a note on terminology is warranted. The terms low resource settings, resource-limited societies, and marginalised societies are used interchangeably in this book and refer to the individuals and communities that typically face the above-mentioned multidimensional challenges.

Design, with its key idea of changing an existing or undesired state into a desired state (Simon 1996), plays a critical role in satisfying unmet or underserved needs of people living in resource-limited societies (Papanek and Fuller 1972). Solutions, designed specifically for these societies, can potentially contribute towards their well-being, creating a significant impact on human development (e.g., Jagtap 2019a). Some examples of such solutions include solutions that provide access to energy, healthcare, water and sanitation services, or some other solutions that support development of these societies or enhance their capabilities (e.g., Jagtap 2019a, b; Aranda-Jan et al. 2016; Ambole et al. 2016). Design of such solutions is carried out by various stakeholders, including governments as their responsibility to deliver community services, NGOs as a form of social service and charitable work, companies as a part of their ongoing efforts to search and enter new markets, and marginalised individuals as a means of sustaining their income generation (Jagtap 2019a; Prahalad 2004).

1.3 Design for Low Resource Settings: Approaches and Concepts

The literature presents several design approaches and concepts aimed at addressing unmet or underserved needs of marginalised people (see Table 1.1). Some of these approaches and concepts are elaborated in the paragraphs that follow.

1.3.1 Appropriate Technology and Design for the Real World

Jagtap (2019a) offers a synthesis of various approaches and concepts about design for low resource settings. In the 1970s, two movements—'appropriate technology' and 'design for the real world'—aimed at designing appropriate solutions to support development of marginalised societies in developing countries, were popularised by Schumacher and Papanek, respectively (Schumacher 1973; Papanek and Fuller 1972). Victor Papanek, an industrial designer, through his book 'Design for Real World', advised designers to design solutions for addressing needs of marginalised people in the Third World. This proposition of using design to alleviate problems of marginalised people was atypical at that time since most of the design work undertaken by designers in the industrialised countries focused on designing products for non-marginalised societies (Amir 2004).

Table 1.1 Various approaches and concepts about design and engineering for low resource settings (synthesised from Jagtap 2019a)

Approaches and concepts	Some references
Appropriate technology, design for the real world	Schumacher (1973), Papanek and Fuller (1972), Jagtap (2019a)
Base of the pyramid	Prahalad and Lieberthal (1998), Prahalad and Hart (1999), Méndez-León et al. (2024)
Subsistence marketplaces	Viswanathan and Sridharan (2009), Viswanathan (2016)
Frugal innovation, social innovation, etc.	Zeschky et al. (2011), Girija et al. (2024)

The appropriate technology movement was popularised by E. F. Schumacher—an economist. The appropriate technology concept formed a basis for his well-recognised book 'Small is Beautiful'. In 1966, Schumacher founded the Intermediate Technology Development Group, which currently functions as Practical Action group. While the appropriate technology concept was articulated by Schumacher, Mahatma Gandhi encouraged developing local technologies suitably tailored to the needs of Indian villages and is thus recognised as founder of appropriate technology movement (Anthony 2000). One of the factors contributing to the popularisation of appropriate technology movement was related to unsuccessful transfer of Western technologies to developing countries during the 1950s and 1960s. This unsuccessful transfer was due to the disparities between socio-cultural and political circumstances between Western and developing countries (Nieusma and Riley 2010). Appropriate technologies were typically placed between advanced technologies from Western countries and traditional tools from villages in developing countries, and were thus recognised as intermediate technologies (Schumacher 1973).

Numerous case studies illustrate the essential design criteria that must be taken into account when developing Appropriate Technologies (e.g., Murphy et al. 2009). These commonly cited design criteria encompass simplicity, affordability, utilisation of locally accessible materials, low energy consumption, potential to promote job creation, and compatibility with socio-cultural contexts (e.g., Jagtap 2019a; Akubue 2000). Although the appropriate technology movement became popular in the 1970s, it faced extensive criticism due to its limited success in achieving sustainable and extensive positive effects in developing countries (Murphy et al. 2009). Several scholars have contended that the prevalence of emotional and philosophical prejudices about the appropriate technology concept can impede a methodical way of designing products that gives priority to user-needs (e.g., Donaldson 2006).

1.3.2   Base of the Pyramid Approach

In the case of appropriate technology and design for the real-world movements of Schumacher and Papanek, the role of non-governmental organisations (NGOs) in designing solutions can be recognised (Jagtap 2019a; Donaldson 2006). On the other hand, in the case of base of the pyramid (BOP) approach, articulated by Prahalad, the role of multinational enterprises (MNEs) in designing solutions for people living in resource-limited societies in developing countries is evident. C. K. Prahalad, along with his collaborators, proposed the notion that MNEs can potentially augment their financial gains while simultaneously addressing widespread poverty (Prahalad and Lieberthal 1998; Prahalad and Hart 1999). A key concept within Prahalad's proposal was the ability of MNEs to enter markets occupied by economically disadvantaged individuals in developing countries by offering them suitable products and services, thereby supporting mutually beneficial outcomes. While prevailing and common notions often consider little or no promise for private sector to alleviate poverty and make profits simultaneously, Prahalad and his collaborators asserted that MNEs can create a win–win situation by satisfying needs of low-income people and making profits. In line with the appropriate technology concept, the BOP approach has also been scrutinised, in particular by Karnani (2006). He argued that for-profit companies can contribute towards alleviating problems of people living in low resource settings by considering them not just as consumers but also as producers and by generating income opportunities for them.

1.3.3   Subsistence Marketplaces

In his synthesis of various approaches and concepts in the field of design for low resource settings, Jagtap (2019a) has included subsistence marketplaces approach devised by Viswanathan and his collaborators (e.g., Viswanathan and Sridharan 2009; Viswanathan 2016). This approach emphasises that markets inhabited by low-income individuals and communities are well-connected networks of individuals. These communities employ social as well as informal channels for various kinds of exchanges, including economic exchanges. As such, the subsistence marketplaces approach proposes to employ a detailed approach aimed at gaining in-depth understanding of behaviours of low-income individuals and their wider context. Subsistence marketplaces approach advocates to learn from these marketplaces rather than considering them as just markets for selling products (Viswanathan et al. 2011). The approach suggests to glean deeper understanding about life situations of people living in low resource settings in the process of designing solutions for them.

1.3.4   Frugal Innovation and Social Innovation

In addition to the above-mentioned approaches such as appropriate technology, BOP, and subsistence marketplaces, in recent years, design of solutions for low-income people is discussed using a variety of names such as grass root innovation (Gupta 2016; Molina-Betancur et al. 2022), jugaad innovation (Radjou et al. 2012; McCausland 2023) and frugal innovation (e.g., Zeschky et al. 2011; Girija et al. 2024). These innovations commonly involve design of cost-effective solutions in a resource-limited setting. Although these solutions may not employ sophisticated technology, they are capable of addressing essential functions (Agnihotri 2015; Jagtap 2019a). Furthermore, strategies using various labels such as community development engineering, design for extreme affordability, engineering for development, humanitarian engineering, social entrepreneurship, and social innovation are recognised in the field of design for low resource settings (e.g., Donaldson 2009; Falcioni 2011; Jagtap 2019a).

1.4 Integrated Design Methodology

Designing solutions to address the needs of marginalised populations necessitates addressing broad range of challenges. These encompass deficiencies in physical infrastructure, the absence of established formal institutions and consistent regulatory frameworks, significant resource scarcity, limited literacy and numeracy levels of marginalised individuals, as well as their low and inconsistent income (Jagtap et al. 2013; UNDP 2008; Prahalad 2012; Jagtap and Larsson 2018). In the context of low resource settings, the process of designing solutions is predominantly motivated by the imperative to address a multitude of limitations and disadvantages. As such, in addition to fulfilling essential functions, these solutions must also satisfy several additional demands emerging from the various shortcomings in these communities. As a consequence, solutions designed to enhance the life quality of people living in marginalised societies are often comprehensive, integrated solutions. For example, these solutions may consist of effective strategies for addressing problems related to constrained distribution networks. These solutions may also necessitate the development of context-specific systems for repairing, maintaining, and recycling the solutions (e.g., Jagtap and Larsson 2013; UNDP 2008). Numerous studies have highlighted the importance of integrated, comprehensive solutions to address the needs of marginalised communities (e.g., Jagtap 2019b; Aranda-Jan et al. 2016).

The subject of designing solutions in the context of resource-limited societies is discussed using a variety of approaches and concepts such as base of the pyramid (BOP), subsistence marketplaces, product service systems (PSS), holistic design for low resource settings, social innovation, social entrepreneurship (e.g., Jagtap 2019a, b; Aranda-Jan et al. 2016; Goyal and Sergi 2015; Nielsen and Samia 2008). Jagtap (2019b) conducted a comprehensive systematic review of an extensive body of literature within this domain, distilling ten guidelines aimed at supporting design of holistic, integrated solutions within this field. The guidelines are derived from recurring findings of the reviewed studies. These guidelines should not be viewed as inflexible or automated processes; instead, they ought to be considered in designing solutions for achieving a positive, intended impact on the lives of marginalised individuals by facilitating the creation of effective solutions (Jagtap 2019b). The ten guidelines are as follows:

1. Knowing context holistically.
2. Codesign in low resource settings.
3. Contextual adaptation.
4. Harnessing local resources and strengths.
5. Context-sensitive training programmes.
6. Integrating income opportunities.
7. Designing tailored awareness plans.

8. Enforcing life cycle needs.
9. Intersectoral collaboration.
10. Customising project management and preventing biases.

Previously, Sect. 1.1 included a design process model consisting of the phases—(1) understanding context and needs, (2) concept exploration and detailing, and (3) implementation. This model presents the design process at a broader level. It is crucial to use the above ten guidelines in relevant phases of the process. These guidelines usefully support pertinent activities in different phases (Jagtap 2021). The guidelines can also offer some recommendations and methods on how design activities in different phases of the process can be accomplished. The ten guidelines help in carrying out various activities in the design process in order to design, develop, and implement solutions aimed at fulfilling unmet or underserved needs people living in resource-limited societies. While some guidelines are more relevant to a specific phase of the design process, some are relevant in all phases (see Fig. 1.2). For example, the guideline 'knowing context holistically' is predominantly applicable in the first phase. On the other hand, the guideline 'codesign in low resource settings' is applicable in all the three phases. Mapping of the ten guidelines over different phases of the design process model results into an integrated design methodology for low resource setting (Jagtap 2021). Figure 1.2 represents this integrated design methodology.

The ten guidelines mapped over the design process provide the basis for the remaining chapters of this book. The remaining chapters present details of these guidelines and offer easy-to-use, flexible, and effective strategies and methods for how these guidelines can be used. The second chapter provides details of the three guidelines—knowing context holistically, contextual adaptation, and harnessing local resources and strengths. The third chapter presents details of the guideline—codesign in low resource settings. While fourth chapter presents details of the three guidelines—context-sensitive training programmes, designing tailored awareness plans, and integrating income opportunities, the fifth chapter explains the remaining three guidelines—enforcing life cycle needs, intersectoral collaboration, and customising project management and preventing biases. Each guideline is explained by providing a case study illustrating the importance and use of the guideline. These case studies are typically based on real-world design projects, covering a broad range of sectors such agriculture, energy, healthcare, housing, etc.

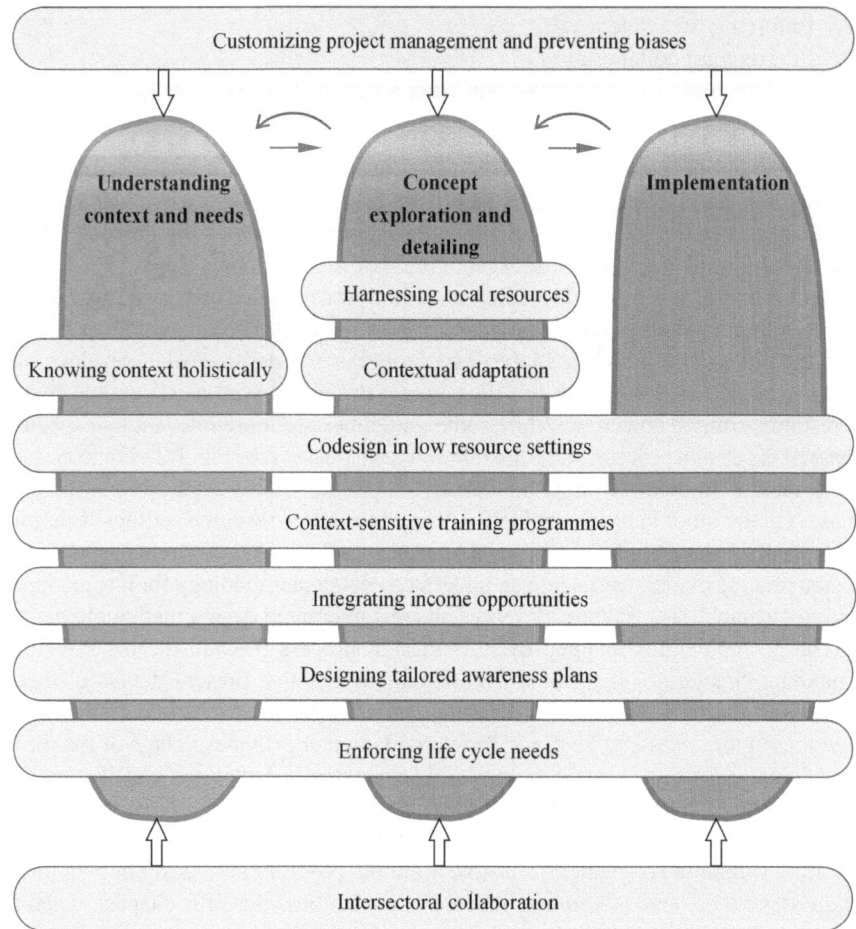

Fig. 1.2 Integrated design methodology for low resource settings (adapted from Jagtap 2021)

References

A. Agnihotri, Low-cost innovation in emerging markets. J. Strateg. Mark. **23**(5), 399–411 (2015)

A. Akubue, Appropriate technology for socioeconomic development in third world countries. J. Technol. Stud. **26**(1), 33–43 (2000)

L.A. Ambole, M. Swilling, M.K. M'Rithaa, Designing for informal contexts: A case study of Enkanini sanitation intervention. Int. J. Des. **10**(3), 75–84 (2016)

S. Amir, Rethinking design policy in the third world. Des. Issues **20**(4), 68–75 (2004)

A. Anthony, Appropriate technology for socioeconomic development in third world countries. J. Technol. Stud. **26**(1), 33–43 (2000)

C. Aranda-Jan, S. Jagtap, J. Moultrie, Towards a framework for holistic contextual design for low-resource settings. Int. J. Des. **10**(3), 43–63 (2016)

N. Cross, *Engineering Design Methods: Strategies for Product Design* (Wiley, 2023)

K. Donaldson, The future of design for development: Three questions. Inform Technol. Int. Dev. **5**(4), 97 (2009)

K.M. Donaldson, Product design in less industrialized economies: Constraints and opportunities in Kenya. Res. Eng. Design **17**(3), 135–155 (2006)

J. Falcioni, Research in extreme affordability. J. Mech. Eng. **133**(5), 6 (2011)

S. Girija, B. Banerji, N. Batra, M. Paruchuru, T. Yeediballi, Making frugal innovations inclusive: A gendered approach. J. Clean. Prod. **434**, 140182 (2024)

S. Goyal, B.S. Sergi, Social entrepreneurship and sustainability: Understanding the context and key characteristics. J. Secur. Sustain. Issues **4**(3) (2015)

A.K. Gupta, *Grassroots innovation: Minds on the margin are not marginal minds* (Random House India, 2016)

S. Jagtap, A. Larsson, P. Kandachar, Design and development of products and services at the base of the pyramid: A review of issues and solutions. Int. J. Sustain. Soc. **5**(3), 207–231 (2013)

S. Jagtap, Design and poverty: A review of contexts, roles of poor people, and methods. Res. Eng. Design **30**, 41–62 (2019a)

S. Jagtap, Key guidelines for designing integrated solutions to support development of marginalised societies. J. Clean. Prod. **219**, 148–165 (2019b)

S. Jagtap, T. Larsson, Design and frugal innovations: Three roles of resource-poor people, in *DS 92: Proceedings of the DESIGN 2018 15th International Design Conference* (2018), pp. 2657–2668

S. Jagtap, A. Larsson, Design of product service systems at the base of the pyramid, in *ICoRD'13* ed. by A. Chakrabarti, R.V. Prakash. Lecture notes in mechanical engineering (Springer, India, 2013), pp. 581–592

S. Jagtap, A. Larsson, V. Hiort, E. Olander, A. Warell, P. Khadilkar, How design process for the base of the pyramid differs from that for the top of the pyramid. Des. Stud. **35**(5), 527–558 (2014a)

S. Jagtap, A. Warell, V. Hiort, D. Motte, A. Larsson, Design methods and factors influencing their uptake in product development companies: A review, in *DS 77: Proceedings of the DESIGN 2014 13th International Design Conference* (2014b)

S. Jagtap, Frugal-IDeM: An integrated methodology for designing frugal innovations in low-resource settings, in *Design for Tomorrow—Volume 2: Proceedings of ICoRD 2021* (Springer Singapore, 2021), pp. 41–51

A. Karnani, *Mirage at the Bottom of the Pyramid: How the Private Sector Can Help Alleviate Poverty (Working Paper no 835)* (William Davidson Institute, University of Michigan, 2006)

V. Kumar, *101 Design Methods: A Structured Approach for Driving Innovation in Your Organization* (Wiley, 2012)

T. McCausland, Reverse innovation, frugal innovation, and Jugaad. Res. Technol. Manag. **66**(1), 68–70 (2023)

E. Méndez-León, R. Díaz-Pichardo, T. Reyes-Carrillo, M. del Rosario Reyes-Santiago, What is unique about sustainable business models for the base of the pyramid? Bus. Strateg. Environ. **33**(3), 2345–2366 (2024)

J.C. Molina-Betancur, A.A. Agudelo-Suárez, E. Martínez-Herrera, Grassroots innovation practices for social transformation of the health and well-being in a self-built settlement in Medellín-Colombia. Health Soc. Care Community **30**(5), 1809–1817 (2022)

H.M. Murphy, E.A. McBean, K. Farahbakhsh, Appropriate technology—A comprehensive approach for water and sanitation in the developing world. Technol. Soc. **31**(2), 158–167 (2009)

C. Nielsen, P.M. Samia, Understanding key factors in social enterprise development of the BOP: A systems approach applied to case studies in the Philippines. J. Consum. Market **25**, 446–454 (2008)

D. Nieusma, D. Riley, Designs on development: Engineering, globalization and social justice. J. Int. Network Eng. Stud. **2**, 29–59 (2010)

G. Pahl, W. Beitz, J. Feldhusen, K.H. Grote, *Engineering Design: A Systematic Approach*, vol. 3 (Springer, London, 1996)

V. Papanek, R.B. Fuller, *Design for the Real World* (Thames and Hudson, London, 1972), p. 22

C.K. Prahalad, S. Hart, *Strategies for the Bottom of the Pyramid: Creating Sustainable Development (Working Paper)*. University of Michigan, Ann Arbor. http://www.bus.tu.ac.th/usr/wai/xm622/conclude%620monsanto/strategies.pdf. Accessed 5 Dec 2016 (1999)

C.K. Prahalad, K. Lieberthal, The end of corporate imperialism. Harvard Bus. Rev. **76**(4), 68–79 (1998)

C.K. Prahalad, *The fortune at the Bottom of the Pyramid: Eradicating Poverty Through Profits* (Wharton School Publishing, Upper Saddle River, NJ, 2004)

C.K. Prahalad, Bottom of the pyramid as a source of breakthrough innovations. J. Prod. Innov. Manag. **29**(1), 6–12 (2012)

N. Radjou, J. Prabhu, S. Ahuja, *Jugaad Innovation: Think Frugal, be Flexible, Generate Breakthrough Growth* (Wiley, New York, 2012)

M. Rivera-Santos, C. Rufín, Global village versus small town: Understanding networks at the Base of the Pyramid. Int. Bus. Rev. **19**(2), 126–139 (2010)

E.F. Schumacher, *Small is Beautiful: Economics as If People Mattered* (Harper and Row, New York, 1973)

H.A. Simon, *The Sciences of the Artificial* (MIT press, Cambridge, MA, 1996)

K.T. Ulrich, S.D. Eppinger, M.C. Yang, *Product Design and Development* (McGraw-hill, 2020)

UNDP, *Creating Value for All: Strategies for Doing Business with the Poor* (United Nations Development Programme, 2008)

M. Viswanathan, S. Sridharan, From subsistence marketplaces to sustainable marketplaces: A bottom-up perspective on the role of business in poverty alleviation. Ivey Bus. J. **73**(2), 1–15 (2009)

M. Viswanathan, A. Yassine, J. Clarke, Sustainable product and market development for subsistence marketplaces: Creating educational initiatives in radically different contexts. J. Prod. Innov. Manag. **28**(4), 558–569 (2011)

M. Viswanathan, *Bottom-up enterprise: insights from subsistence marketplaces* (2016)

M. Zeschky, B. Widenmayer, O. Gassmann, Frugal innovation in emerging markets. Res-Technol. Manag. **54**(4), 38–45 (2011)

Chapter 2
Context: Holistic Understanding and Adaptation

Abstract This chapter presents details of three key guidelines that can potentially support the process of designing, developing, and implementing effective solutions for the betterment of resource-constrained population. Firstly, it presents detailed information on the guideline about developing holistic understanding of the local context. Secondly, the chapter highlights the importance of the guideline about adapting solutions to the specific circumstances and conditions in the local context. Finally, it presents details of the guideline about building on strengths within the local context, suggesting that leveraging socio-cultural assets and employing existing resources in resource-constrained communities can lead to more impactful solutions. For each of these three guidelines, the chapter presents a case study, while offering recommendations on how the guidelines can be implemented in a project.

2.1 Knowing Context Holistically

Developing comprehensive understanding of the wider context of low resource settings and patterns of daily lives of marginalised individuals is an essential part of design and engineering in this field (Jagtap 2019b).

2.1.1 Why to Know Local Context Holistically

Marginalised communities encounter numerous multifaceted challenges. Scholars have attempted to classify the problems faced by marginalised communities into various categories (Jagtap 2019b). For example, Aranda-Jan et al. (2016) have classified these challenges into categories such as individual, infrastructural, institutional, and technological issues. On the other hand, Nakata and Weidner (2012) have categorised these problems into the following types: physical issues, economic issues, psychosocial issues, and knowledge-related issues. These categories highlight the multifaceted nature of the problems encountered by marginalised individuals. These problems are interconnected and exist at various levels, creating

S. Jagtap, *Design and Engineering for Low Resource Settings*,
SpringerBriefs in Applied Sciences and Technology,
https://doi.org/10.1007/978-3-031-66156-3_2

complex challenges for these communities. The studies suggest that addressing the issues faced by marginalised communities requires a comprehensive and holistic approach. To meet the unfulfilled needs of these communities, it is imperative to tackle numerous value chain gaps prevalent in the local context. This is supported by studies conducted by Devisscher and Mont (2008) and Jagtap et al. (2013). Previous studies have demonstrated that effective solutions for marginalised communities require all-inclusive comprehension of the context. This necessitates considering broader institutional, socio-cultural, and structural issues that exist within these communities (e.g., Aranda-Jan et al. 2016; Jagtap and Larsson 2013; López et al. 2017; Letaifa and Reynoso 2015). According to the studies carried out in Brazil's transportation and water sectors (Sousa-Zomer and Miguel 2016a; Sousa-Zomer and Cauchick-Miguel 2017), solutions that seek to assist marginalised communities in their social and human development should be based on thorough and integrated comprehension of the specific context. These studies indicate that a holistic approach is essential for achieving success in supporting the development of disadvantaged societies. In a similar fashion, Nakata and Weidner (2012) highlight the importance of understanding unique challenges within a specific context. They substantiate this claim by referencing several design examples from a variety of sectors in developing nations.

Several other studies have proposed that adopting a bottom-up approach to learning from marginalised individuals can be beneficial for gaining a comprehensive understanding of the context of interest (Whitney 2010; Viswanathan 2017). In general, these studies argue that anchoring solutions in the daily life circumstances of underprivileged individuals is important for the persistent uptake and utilisation of the solutions (Whitney 2010). According to Viswanathan (2010), gaining a nuanced understanding of marginalised communities requires a holistic immersion in their context because there can be significant differences across various contexts. Such an understanding is essential for outsiders to overcome their unfamiliarity with low resource settings and establish personal connections with individuals from these settings. According to Viswanathan (2016), adopting a bottom-up learning approach that considers the broader context in which marginalised individuals live can provide valuable insights into various aspects of their lives. These insights may include their living conditions, social networks, methods of producing things, thought processes, and their approaches to environmental challenges.

2.1.2 How to Know Local Context Holistically

Several studies have offered methods in order to develop holistic and comprehensive understanding of marginalised communities (Jagtap 2019b). These methods are as follows.

Holistic design framework: Aranda-Jan et al. (2016) have developed a design framework intended to gather information on marginalised communities in the early phases

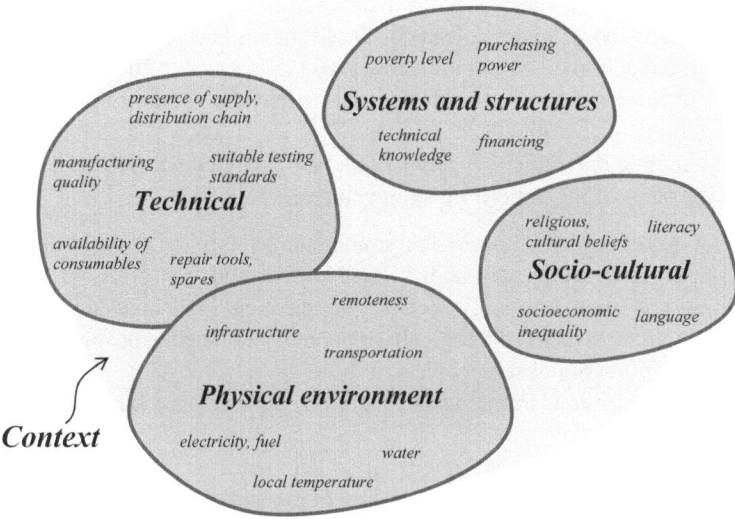

Fig. 2.1 Holistic design framework (adapted from Aranda-Jan et al. 2016)

of the design process. This comprehensive framework includes four primary categories of factors and subcategories in each of these primary categories (see Fig. 2.1). The authors argue that utilising this holistic approach supported by their holistic design framework may aid designers in acquiring a more profound comprehension of the requirements and difficulties confronted by marginalised communities.

Framework of productivity and transactional constraints: London et al. (2010) conducted an analysis of 64 case studies to develop a framework that identifies various constraints faced by marginalised individuals in managing their businesses. The ability of marginalised producers to create value is impacted by two main categories of limitations within the framework. The first category includes productivity constraints that hinder their value creation. The second category encompasses transactional constraints that inhibit their capacity to benefit from selling their products in both local and non-local markets. The framework is designed to aid in the development of a complete and thorough understanding of the issues and challenges that are confronted by marginalised producers. The framework divides the subject matter into the above two main categories, each with its own subcategories. By exploring these categories and subcategories, the framework allows for a more comprehensive approach for identifying and addressing the various constraints faced by marginalised producers.

The POEMS framework: The POEMS framework—People, Objects, Environments, Messages, and Services—was employed by Whitney and Kelkar in 2004 for developing solutions for individuals residing in slums located in Mumbai. This framework is a promising alternative for gaining a comprehensive understanding of the context being targeted.

Bottom-up approach: Various studies have been conducted to collect comprehensive information about the target context. These studies have developed frameworks that support collection of information about the target context. Viswanathan (2016) proposed an interesting and useful approach called the 'bottom-up approach' to design solutions for marginalised societies. This approach is particularly useful in creating solutions that satisfy specific needs and requirements of these societies. Some of the steps of the bottom-up approach are as follows:

1. Virtual immersion: Conduct virtual immersion by employing multiple methods and media such as textual, audio-visual media, and transcripts of interviews with marginalised people. This offers participants (e.g., 'outsiders' unfamiliar with life circumstances of marginalised people) simulated exposure to the living conditions of marginalised people.
2. Engage in emersion: Participants reflect on the insights gleaned from virtual immersion.
3. Actual immersion: Participants engage in field research in the target context to gain direct experience and understanding of the situation.
4. Using insights: Participants use the insights gained from virtual immersion, emersion, and actual immersion to gain a holistic understanding of the target context.

In addition to the above methods, there are traditional methods that can be used by both organisations and individuals to gather valuable insights. The methods include conducting interviews, shadowing, and field observations, engaging in role-playing exercises, and using ethnography to understand cultural practices (Jagtap 2019b).

It is crucial to holistically gain insights into the local context in the initial phase of the design process (Jagtap 2021). This is illustrated in Fig. 2.2, together with how some methods and strategies that can assist in knowing the local context holistically.

Box 2.1 Knowing Context Holistically—A Case Example

The effectiveness of solutions designed for marginalised communities is heavily influenced by the extent to which they are founded on a comprehensive and nuanced understanding of the specific context (Jagtap 2019b). In the state of Maharashtra in India, a project named the 'Warana Wired Village Project' was implemented in a rural district. This initiative aimed to provide internet access to approximately 40,000 farmers in the area, with the goal of enabling them to access information on crop cultivation and market prices. The project's purpose was to assist the farmers in selling their produce competitively (Toyama et al. 2007). The assessment of the project impact was carried out by a collaborative team of researchers from the University of Berkeley, the Microsoft Research India, and the London School of Economics. The team employed a range of research methodologies including surveys, observations, interviews, and ethnography. The impact assessment revealed that the Internet kiosks implemented in the target context did not fulfil their intended purpose.

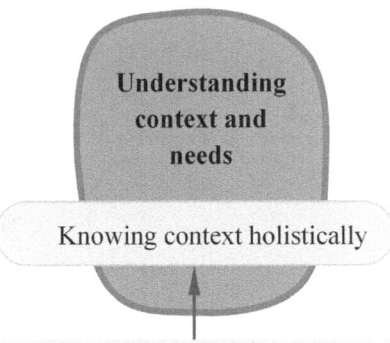

How to know local context holistically

- Holistic design framework
- Framework of productivity and transactional constraints
- POEMS framework
- Bottom-up approach
- Conventional methods such as observations

Fig. 2.2 Design process phase and methods for knowing context holistically

The reasons behind this included weak access to electricity, which limited the functionality of the kiosks, and a lack of familiarity with the technology among the farmers. Additionally, poor maintenance of the kiosks contributed to their ineffective use. These findings highlight the fact that the project was not adequately embedded in the holistic understanding of the target context, leading to a low level of success in achieving its goals. In essence, the implementation of the project failed to address key contextual factors, which ultimately hindered its impact and led to suboptimal outcomes. To address these challenges, a project called 'Warana Unwired' was developed. This initiative was carefully designed with a deep understanding of the local context and aimed to provide solutions that are both accessible and effective. One of the key features of the 'Warana Unwired' project was its use of mobile phones as a primary communication tool. The project team recognised that farmers in the region were already familiar with mobile phones and could use them with ease, regardless of their location or time of day. By using this technology, the project was able to create a communication system that was accessible to the farmers in the area. Compared to the previous project that relied on kiosk computers, the 'Warana Unwired' project was perceived as more useful by the farmers (Jagtap 2019b). Figure 2.3 illustrates the above case example.

Fig. 2.3 Illustration of the
'Warana Unwired' case
example

2.2   Contextual Adaptation

Adapting solutions to the local specificities and living conditions of people in the target context is important for their continued adoption and use by the marginalised people (Jagtap 2019b).

2.2.1   Why to Adapt Solutions to Local Conditions

Numerous distinct limitations and deprivations, commonly encountered in disadvantaged communities, are absent in non-disadvantaged settings where the accessibility of vendors, promotional channels, distribution systems, power supply, financial channels, and transportation systems can be presumed (e.g., Jagtap 2019b). Numerous deprivations and specific socio-cultural attributes of disadvantaged communities necessitate careful adaptation of solutions to regional nuances and circumstances of these communities (e.g., Gebauer et al. 2017; Jagtap et al. 2013). According to Prahalad (2012), the long-term and sustainable impact of a designed solution on a marginalised community depends on how well it is adapted to local conditions.

In a study conducted by Ernst et al. (2015), a total of 103 projects designed for marginalised societies across multiple countries and design sectors were analysed. The study, which was conducted on a large scale and covered various regions, offers convincing evidence that adapting solutions to the specificities and conditions of local communities is vital for their optimal performance. The findings of this study suggest that a design approach that prioritises local considerations can result in better outcomes for marginalised individuals and communities. This requires developing solutions that not only meet needs of these communities, but also address the larger systemic issues they face. The United Nations Development Programme (UNDP) conducted a study called 'Growing Inclusive Markets' that focused on various sectors such as agriculture, energy, healthcare, transportation, water, and sanitation in developing countries (UNDP 2008). The study emphasised the importance of tailoring solutions to the local context for the sustained adoption and use of developed solutions by marginalised individuals.

In addition to the above-mentioned large-scale studies conducted by Ernst et al. (2015) and UNDP (2008), numerous studies have highlighted the importance of adapting solutions to the local context for positive impact on marginalised communities. It is widely recognised that context-specific solutions have a greater potential to meet the unique needs and challenges of marginalised communities. For instance, Sousa-Zomer and Miguel (2016b) conducted an analysis on a reverse osmosis (RO) water system in Brazil aimed at providing purified water to the people living in the southern region. Their findings suggest that successful performance of the solution is highly dependent on the adaptation of the entire system to address different constraints and requirements specific to the target context. Another example is about lighting solutions. D.light has developed innovative lighting solutions for rural households and off-grid schools, utilising solar and LED technology (Emili et al. 2016). These lighting solutions were specifically designed for rural households and off-grid schools by adjusting their products, financial channels, and distribution strategies to meet the unique needs of each context.

There is a significant amount of variability observed among developing countries and across rural and urban regions (e.g., Dahana et al. 2018; Jagtap 2019a). For example, there are dissimilarities in profession and spatial distribution among people residing in rural and urban regions. In contrast to rural areas where underprivileged groups are scattered geographically, urban slums have a higher population density (Johnson 2007). Furthermore, there exist divergences in the social connections of marginalised individuals in rural and urban areas (Sridhar 2015). The development of solutions that meet the diverse needs of communities is a key challenge for designers. Whitney and Kelkar's (2004) provide an excellent example of the importance of considering context in design. Their design solution aimed to address the lack of access to clean water for slum dwellers in Mumbai, India. The proposed solution was developed with the living conditions of slum dwellers and the broader context of slums in the city in mind. This approach highlights the crucial role of contextualisation in the development of effective and sustainable solutions for communities in both rural and urban areas.

2.2.2 How to Adapt Solutions

According to Jagtap et al. (2013), it is crucial to obtain a comprehensive understanding of marginalised communities. Furthermore, it is equally significant to utilise this understanding for adapting solutions to the living circumstances of marginalised individuals and to the wider context in which they reside (Jagtap 2019b).

Using insights from bottom-up learning: According to Aranda-Jan et al. (2016), solutions need to be based on insights acquired through bottom-up learning from marginalised communities. Additionally, ensuring the suitability of the solutions for the target context can enhance solution adaptability.

Adaptation of entire solution: It is important to adapt various components of a solution in a synergistic manner to contextualise the entire solution for marginalised communities (e.g., Devisscher and Mont 2008). This includes adapting entire solution, including its various aspects such as products, services, supporting infrastructure, and requisite networks. This approach can help ensure that solutions are tailored to the specific needs and circumstances of marginalised communities.

Implementation of solution attributes: One of the recommendations to adapt solutions to the context of marginalised communities is to consider various solution attributes (Nakata and Weidner 2012). Some of these attributes are as follows:

1. Affordability: It is a crucial factor in ensuring that solutions are accessible to lower-income groups.
2. Interpersonal marketing: This involves leveraging personal connections to promote and distribute solutions.
3. Visual comprehensibility: Visually comprehensible solutions refer to solutions that are easily understandable and usable, regardless of literacy levels or language barriers.
4. Atomised distribution: This involves the use of small-scale distribution networks for greater reach.
5. Flexible payment schemes: Flexible payment options make it easier for marginalised communities to access and afford solutions.

Figure 2.4 offers an illustration of the adaptation of solutions to the local context by considering the above attributes. By incorporating these attributes into solutions, they can better meet the needs of marginalised communities and have a greater chance of success.

Using insights gleaned from the task of knowing context holistically in the initial phase of the design process, to adapt solutions to the local context needs to be undertaken in the phase 'concept exploration and detailing' (Jagtap 2021). Figure 2.5 shows the applicability of adapting solutions to the local context in the phase 'concept

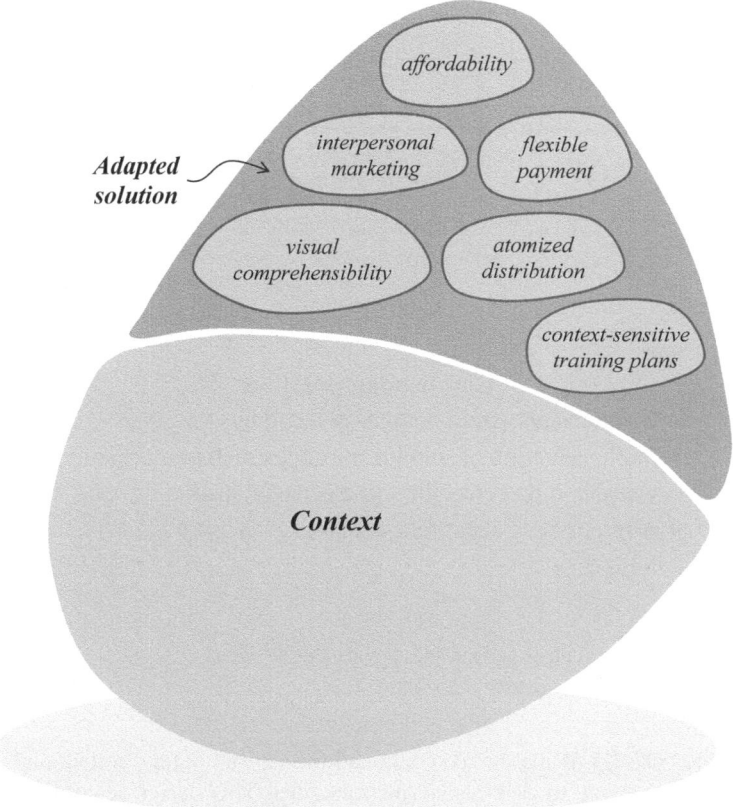

Fig. 2.4 Consideration of various attributes in adapting solutions to the local context

exploration and detailing', together with methods and strategies that can be used in contextual adaptation of solutions.

Box 2.2 Contextual Adaptation—A Case Example

Whitney's (2010) research investigates the design of a refrigeration system called 'Chotukool' which was specifically developed to serve the refrigeration needs of low-income individuals residing in rural areas of India. The study emphasises the significance of adapting solutions to specific contexts, considering the socio-cultural characteristics, literacy levels, and life circumstances of the target audience. The Chotukool's design was specifically tailored to the local context, carefully considering the distinctive obstacles encountered by rural communities in India. As an illustration, the products were designed to be easily transportable and lightweight, employing solid-state induction plates instead of compressors to endure voltage fluctuations plus scorching and dusty

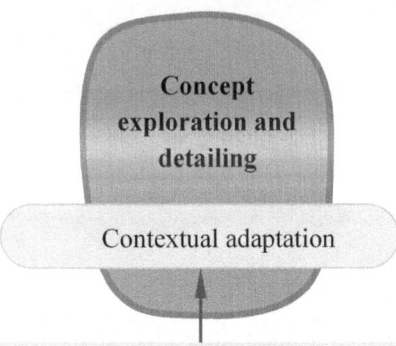

Fig. 2.5 Design process phase and methods for adapting solutions

weather. The lid of the device contained the entire cooling system and was easily replaceable in case of any malfunctions. This design feature ensured that any problems with the cooling system could be quickly and easily fixed. This supported unskilled rural entrepreneurs to perform basic maintenance and repair services. Figure 2.6 illustrates the Chotukool refrigeration solution. The business model of the Chotukool solution was also tailored to the local context. It was distributed through an entrepreneurial network of local people, creating livelihood opportunities. This approach allowed the solution not only to meet the refrigeration needs of low-income consumers but also to have a positive impact on the local economy. Overall, Whitney's case study emphasises the importance of taking a context-specific approach to the design of solutions. By considering the unique challenges and circumstances faced by the target audience, it is possible to create solutions that are effective and sustainable, while also generating livelihood opportunities and creating a positive impact on the local economy (Jagtap 2019b).

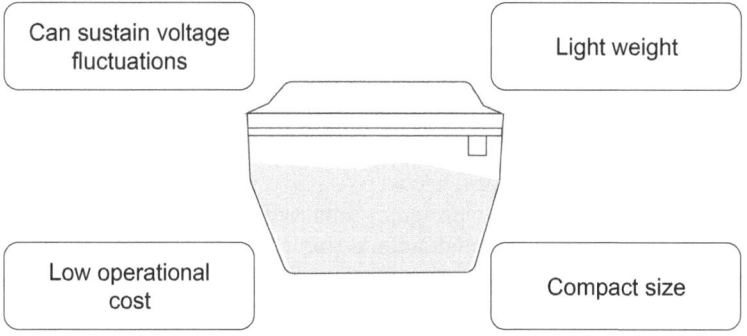

Can sustain voltage fluctuations

Light weight

Low operational cost

Compact size

Fig. 2.6 Illustration of the 'Chotukool' refrigeration solution

2.3 Harnessing Local Resources and Strengths

Solutions for marginalised communities have greater impact when they are designed by leveraging strengths of marginalised people and resources available in the target context (Jagtap 2019b).

2.3.1 Why to Build on Local Resources and Strengths

Individuals living in marginalised communities have developed a set of survival skills that enable them to deal with the challenges of living in conditions where resources are limited (Jagtap 2019b). These skills may include resourcefulness, innovation, and resilience when faced with adversity. It is important to recognise that despite the difficult circumstances they face, people in these communities possess valuable knowledge and experience. Marginalised communities are often associated with informal economies, which emerge due to limited access to conventional economic and political resources, or due to unstable and unpredictable conditions, such as food insecurity (e.g., Jagtap 2019b). Due to their distinctive situations, marginalised groups often display a range of unique qualities. They may exhibit behaviours and attitudes that differ from those of more privileged groups.

Culture of sharing products and resources: Marginalised communities tend to have a culture of sharing products and resources (e.g., Jagtap and Larsson 2013). This is different from developed countries, where individual autonomy and well-being are more emphasised. In marginalised societies, the focus is on collective needs and group goals. This highlights the significance of culture in influencing attitudes and behaviours towards resource sharing.

Trusting relationships: According to Viswanathan (2010), marginalised communities often prioritise building trust among their members. This is exemplified in their approach to marketplace negotiations, where the relationship between parties is valued over the transaction itself. This highlights the significance of interpersonal connections and trusting relationships in such communities.

Cultural norms: Marginalised societies are often guided by their traditional views and established cultural norms (e.g., Rivera-Santos and Rufín 2010). These norms may be based on religion, race, kinship, or other factors, and they tend to hold greater influence than formal institutional agencies.

A more effective approach to designing solutions for marginalised communities is to identify and utilise their social-cultural strengths rather than solely addressing their weaknesses (e.g., Jagtap 2019b). By taking into account the positive aspects of these communities, solutions can be tailored to better fit their needs and achieve greater success. This perspective suggests that a strengths-based approach may lead to superior outcomes in the design and implementation of solutions for marginalised communities. According to Jagtap et al. (2013), it is crucial to consider social networks and patterns in marginalised communities when designing solutions for their betterment. This involves a focus on relationships, welfare, and social fabric of these communities. According to the study conducted by London and Hart (2004) that spanned three years, the success of solutions tailored for marginalised individuals is influenced by various factors. The researchers analysed these factors during the study period to gain a better understanding of the contributing elements. Their study employed multiple data sources such as historical records, primary and pre-existing case studies, and interviews to conduct a thorough examination of 24 cases. The success of developed solutions is found to be related to the utilisation of the strengths of targeted marginalised communities, according to the outcomes of this comprehensive investigation. These findings highlight the significance of leveraging the abilities and resources of these societies for the success of designed solutions. Furthermore, according to UNDP's (2008) large-scale study, which investigated numerous solutions, leveraging the strengths of marginalised communities is a crucial design strategy in this field.

Whilst many studies suggest that it is beneficial to utilise the pre-existing socio-cultural strengths of marginalised communities, several studies have shown that it is advantageous to build upon the products, infrastructure, and resources already present within these societies (e.g., Jagtap 2019b). Ernst et al. (2015) carried out an empirical investigation, scrutinising numerous projects in low resource settings. The results of their study uncovered that bricolage, which refers to the novel and creative use of scarce resources to generate solutions for problem-solving and identifying opportunities, has a favourable impact on the development of solutions for marginalised societies. Utilising available resources within a marginalised community has the potential to decrease the costs associated with materials and products, resulting in more cost-effective solutions (e.g., Prahalad 2004). This approach can also improve the overall affordability of the designed solutions. Available products in marginalised societies can be employed in designing solutions for fulfilling their relevant needs. For instance, widely available mobile phones can be utilised as a means of addressing educational and informational requirements of the marginalised population (e.g., Viswanathan 2010).

2.3.2 How to Build on Local Resources and Strengths

Identification of strengths: In order to leverage strengths of marginalised communities, it is crucial to initially identify and acknowledge these strengths (Jagtap 2019b). Subsequently, these strengths can be employed in designing solutions for empowerment and upliftment of these communities.

Social embeddedness to identify strengths: Numerous studies suggest that social embeddedness plays a vital role in acquiring a comprehensive understanding of socio-cultural and other assets in low resource settings (Jagtap 2019b). It supports gaining insights into available infrastructure, materials, and existing products within the target context of marginalised communities (e.g., Ausrød 2017; Sousa-Zomer and Miguel 2016a, 2016b). Additionally, it facilitates efficient resource mobilisation. In marginalised communities, being locally embedded can provide valuable insights into social patterns, individual interactions, and the development of trust-based relationships (Jagtap 2019b). By being deeply connected to the community, individuals can gain a nuanced understanding of the complex dynamics that exist within it. This understanding can be used to glean insights into the social fabric of the community and the ways in which individuals interact with one another. Overall, local embeddedness can be an important factor in gaining a deeper understanding of social dynamics and strengths that exist within these communities.

Creatively building on strengths: According to some studies, one possible approach to designing solutions that takes advantage of identified strengths involves creatively utilising sharing culture, available products, and social control. This approach can be applied throughout the entire life cycle of the solution to address various functions (Jagtap 2019b). Figure 2.7 offers an illustration of building on strengths of resource-limited societies.

Building on rather than replacing strengths: Incorporating existing strengths of marginalised communities, such as informality and local social networks, is critical

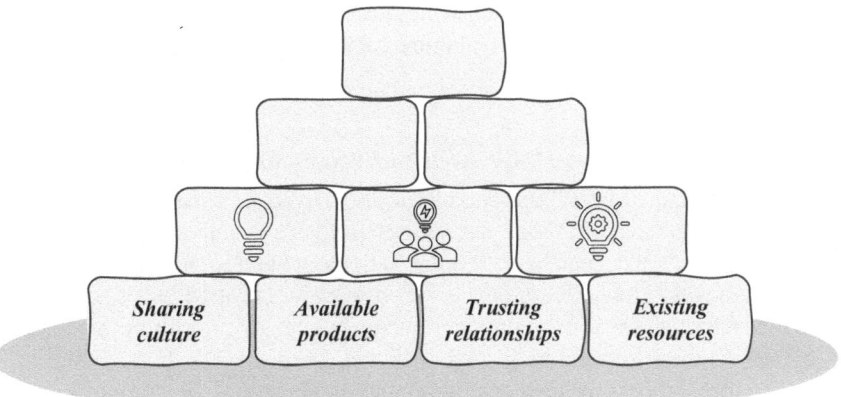

Fig. 2.7 Building on strengths and resources available in resource-limited societies

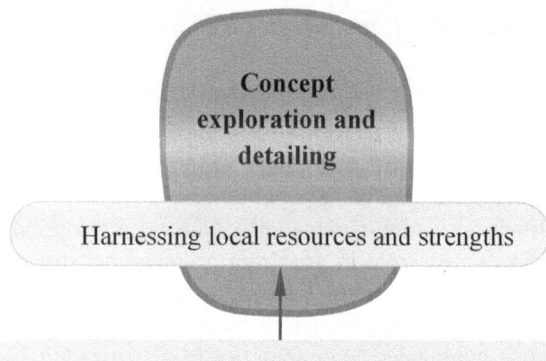

Fig. 2.8 Design process phase and methods for harnessing local resources and strengths

for ensuring the adoption of designed solutions. Rather than imposing dominant approaches from non-marginalised settings, leveraging these strengths can lead to sustainable solutions that are tailored to the specific needs of these communities (e.g., Jagtap et al. 2013). By recognising and building on the assets of marginalised communities, designers can create solutions that better serve their needs and are more likely to be embraced over the long term.

The activities of harnessing strengths and resources available in low resource settings need to be mainly undertaken in the phase 'concept exploration and detailing' of the design process (Jagtap 2021). Figure 2.8 illustrates this aspect, together with related methods and strategies.

Box 2.3 Harnessing Local Resources and Strengths—A Case Example

Devisscher and Mont (2008) conducted an analysis of a solution aimed at assisting low-income coffee producers from Bolivia's Yugnas region. The analysis was based on data gathered through various techniques such as interviews and site visits. The coffee producers used to process coffee in a basic way, individually, before adopting the designed solution. As a result of this, their coffee had weak demand in the global market. Despite some shortcomings in their coffee processing methods, the coffee producers of Bolivia possessed several socio-cultural strengths. In various urban and rural regions of the country, there exists a culture of self-organisation among the people, who work together to establish and manage businesses as well as address community needs. These strengths can be considered as an asset to enhance the coffee production industry in Bolivia, by empowering local communities and facilitating their participation in the value chain. Collectivism prioritises the significance of group unity and cohesion. It is a deeply rooted concept in the rural communities of Bolivia, where the cultural value of collectivism is seen in the collaborative efforts of community members towards common objectives such as agricultural production and formation of community-based organisations. In order to assist coffee farmers in the Yungas region, the Director of Central of Coffee Cooperatives (CENCOOP) devised a strategy that capitalises on the area's socio-cultural assets. This approach was developed to aid coffee producers in the region. CENCOOP developed a joint processing system that comprises systems such as fermentation tanks, pulping machines, and cleaning and drying equipment. The system is intended to process coffee provided by all cooperative members. The shared processing system ensures uniformity and consistency in coffee processing. CENCOOP has taken significant steps to extend their support to the coffee production industry. CENCOOP implemented a range of better production techniques that have helped to significantly enhance the quality of coffee. Additionally, they have also undertaken an initiative to revive old coffee plants. CENCOOP has also introduced a 'Fair Trade' initiative. Due to the utilisation of the communities' several strengths, CENCOOP could support exporting nearly all the coffee it produces (Jagtap 2019b). Figure 2.9 provides an illustration of some aspects of this case example.

Fig. 2.9 Illustration of some aspects of the solution aimed at assisting low-income coffee producers

References

C. Aranda-Jan, S. Jagtap, J. Moultrie, Towards a framework for holistic contextual design for low-resource settings. Int. J. Des. **10**(3), 43–63 (2016)

V.L. Ausrød, It takes two to tango: Mobilizing strategic, ordinary, and weak resources at the base of the pyramid. J. Strat. Market. 1–23 (2017)

W.D. Dahana, T. Kobayashi, A. Ebisuya, Empirical study of heterogeneous behavior at the base of the pyramid: The influence of demographic and psychographic factors. J. Int. Consumer Market. 1–19 (2018)

T. Devisscher, O. Mont, An analysis of a product service system in Bolivia: Coffee in Yungas. Int. J. Innov. Sustain. Dev. **3**(3–4), 262–284 (2008)

S. Emili, F. Ceschin, D. Harrison, Product–service system applied to distributed renewable energy: A classification system, 15 archetypal models and a strategic design tool. Energy Sustain. Dev. **32**, 71–98 (2016)

H. Ernst, H.N. Kahle, A. Dubiel, J. Prabhu, M. Subramaniam, The antecedents and consequences of affordable value innovations for emerging markets. J. Prod. Innov. Manag.innov. Manag. **32**(1), 65–79 (2015)

H. Gebauer, M. Haldimann, C.J. Saul, Business model innovations for overcoming barriers in the base-of-the-pyramid market. Ind. Innov.innov. **24**(5), 543–568 (2017)

S. Jagtap, A. Larsson, P. Kandachar, Design and development of products and services at the base of the pyramid: A review of issues and solutions. Int. J. Sustain. Soc. **5**(3), 207–231 (2013)

S. Jagtap, A. Larsson, Design of product service systems at the base of the pyramid, in *ICoRD'13* ed. by A. Chakrabarti, R.V. Prakash. Lecture Notes in Mechanical Engineering (Springer India, 2013) pp. 581–92

S. Jagtap, Design and poverty: A review of contexts, roles of poor people, and methods. Res. Eng. Design **30**, 41–62 (2019a)

S. Jagtap, Key guidelines for designing integrated solutions to support development of marginalised societies. J. Clean. Prod. **219**, 148–165 (2019b)

S. Jagtap, Frugal-IDeM: An integrated methodology for designing frugal innovations in low-resource settings, in *Design for Tomorrow—Volume 2: Proceedings of ICoRD 2021* (Springer Singapore, 2021), pp. 41–51

S. Johnson, SC Johnson builds business at the base of the pyramid. Glob. Bus. Organ. Excell.excell. **26**(6), 6–17 (2007)

S. Letaifa, J. Reynoso, Toward a service ecosystem perspective at the base of the pyramid. J. Serv. Manag.manag. **26**(5), 684–705 (2015)

T. London, S.L. Hart, Reinventing strategies for emerging markets: Beyond the transnational model. J. Int. Bus. Stud. **35**(5), 350–370 (2004)

T. London, R. Anupindi, S. Sheth, Creating mutual value: Lessons learned from ventures serving base of the pyramid producers. J. Bus. Res. **63**(6), 582–594 (2010)

A.M. López, F. Musonda, T. Sakao, N. Kebir, Lessons learnt from designing PSS for base of pyramid. Procedia CIRP **61**, 623–628 (2017)

C. Nakata, K. Weidner, Enhancing new product adoption at the base of the pyramid: A contextualized model. J. Prod. Innov. Manag.innov. Manag. **29**(1), 21–32 (2012)

C.K. Prahalad, *The fortune at the bottom of the pyramid: Eradicating poverty through profits* (Wharton School Publishing, Upper Saddle River, NJ, 2004)

C.K. Prahalad, Bottom of the pyramid as a source of breakthrough innovations. J. Prod. Innov. Manag.innov. Manag. **29**(1), 6–12 (2012)

M. Rivera-Santos, C. Rufín, Global village versus small town: Understanding networks at the Base of the Pyramid. Int. Bus. Rev. **19**(2), 126–139 (2010)

T.T. Sousa-Zomer, P.A. Cauchick-Miguel, Exploring business model innovation for sustainability: An investigation of two product-service systems. Total Quality Manag. Busi. Excellence 1–19 (2017)

T.T. Sousa-Zomer, P.A.C. Miguel, Exploring the critical factors for sustainable product-service systems implementation and diffusion in developing countries: An analysis of two PSS cases in Brazil. Procedia CIRP **47**, 454–459 (2016a)

T.T. Sousa-Zomer, P.A.C. Miguel, Sustainable business models as an innovation strategy in the water sector: An empirical investigation of a sustainable product-service system. J. Clean. Prod. **171**, S119–S129 (2016b)

K.S. Sridhar, Is urban poverty more challenging than rural poverty? A Review. Environment and Urbanization Asia **6**(2), 95–108 (2015)

Toyama et al., *Review of research on rural PC kiosks.* Microsoft Research India (2007)

UNDP, *Creating Value for All: Strategies for Doing Business with the Poor.* United Nations Development Programme (2008)

M. Viswanathan, *A micro-level approach to understanding BoP markets. Next generation business strategies for the base of the pyramid: New approaches for building mutual value* (FT Press, Upper Saddle River, 2010)

M. Viswanathan, *Bottom-up enterprise: Insights from subsistence marketplaces.* Madhu Viswanathan (2016)

M. Viswanathan, What the subsistence marketplaces stream is really about: Beginning with micro-level understanding and being bottom-up. J. Mark. Manag.manag. **33**(17–18), 1570–1584 (2017)

P. Whitney, Reframing design for the base of the pyramid. Next generation business strategies for the base of the pyramid: New approaches for building mutual value. Upper Saddle River, New Jersey, 165–192 (2010)

P. Whitney, A. Kelkar, Designing for the base of the pyramid. Des. Manage. Rev. **15**(4), 41–47 (2004)

Chapter 3
Codesign in Low Resource Settings

Abstract This chapter explores the importance of codesigning with people living in resource-limited areas. It outlines several advantages of codesign with these people, focusing on improvements to the design process and enhancement of social and human development of these communities. Based on findings gleaned from extant studies, the chapter offers recommendations for how organisations can codesign with these communities, including allocating resources, choosing appropriate codesign methods, and identifying and addressing barriers that may hinder codesign activities. By presenting these findings, the chapter can support organisations with strategies to ensure inclusive and effective codesign with resource-poor populations.

3.1 Codesign in Low Resource Settings

Codesigning with people living in resource-limited societies has several advantages, aiding many different activities in the process of designing, developing, and implementing solutions and supporting acceptance and adoption of designed solutions (Jagtap 2019b).

3.1.1 Why Codesign is Needed in Low Resource Settings

Table 3.1 shows outcomes and advantages of codesign activities with resource-poor people. Of these outcomes and advantages, five relate to improvement of codesign process and the remaining five relate to social and human development of resource-poor communities. The sections that follow present details of these outcomes and advantages.

© The Author(s), under exclusive license to Springer Nature Switzerland AG 2024
S. Jagtap, *Design and Engineering for Low Resource Settings*,
SpringerBriefs in Applied Sciences and Technology,
https://doi.org/10.1007/978-3-031-66156-3_3

Table 3.1 Advantages of codesigning with people living in low resource settings (e.g., Jagtap 2022a)

Category	Advantages of codesign
Codesign process improvement	• Rich contextual understanding • Identification of problems and development of solutions • Collaborative knowledge generation • Assimilation of resources of various partners • Social integration
Development of low resource settings	• Empowerment of resource-poor people • Skill development • Income and livelihood generation • Gender equity • Acceptance and use of solutions

3.1.1.1   Codesign Process Improvement

The outcomes related to improvement of codesign process are about gaining rich contextual understanding, identifying problems and developing solutions, collaborative generation of knowledge, assimilation of resources of various partners, and social integration (e.g., Jagtap 2022a).

Rich contextual understanding: Numerous studies, undertaken in several countries and many different sectors such as energy, agriculture, healthcare, ICT, etc., have revealed advantages of codesigning with marginalised people for designers (e.g., Jagtap 2019b; Brugmann and Prahalad 2007; De Silva et al. 2020; Jagtap and Larsson 2018; Corsini et al. 2022). One of such key advantages is that it helps designers in gaining rich insights into the local context, allowing them to gain holistic, comprehensive understanding of the context of marginalised communities. Codesign helps designers in understanding various constraints, deprivations as well as strengths in marginalised communities (e.g., Jagtap 2019a; Jagtap and Larsson 2019). Codesign also supports them in gaining rich insights into the ways in which marginalised people buy products, perform their entrepreneurial activities, or produce artefacts. It also supports designers in understanding social and economic interactions in marginalised communities, their different types of beliefs, etc.

Identification of problems and development of solutions: Codesigning with marginalised people support formulation of requirements and development of solutions to address the requirements (Jagtap 2022b). For example, it supports designers in formulating different types of requirements associated with social, cultural, economic, and technological aspects. This identification of different types of requirements helps designers in generating and developing appropriate solutions to fulfil needs of marginalised people and in adapting solutions to the local context (e.g., Jagtap 2022a).

Collaborative knowledge generation: Collaborative generation of knowledge is one of the outcomes of codesigning solutions with people living in resource-constrained communities (e.g., Jagtap 2022a).

Assimilation of resources of various partners: Partners involved in cocreating solutions with marginalised communities are typically experts in some areas. For instance, NGOs, due to their social embeddedness, are trusted locally and can be knowledgeable about local issues and resources. Formally trained designers and companies can bring their technical expertise, design knowledge, their networks in global or country-level markets as well as monetary assets. As such, cocreation supports integration of various assets, skills and knowledge of involved partners. This in turn helps in addressing a wide range of constraints and scarcities in resource-constrained communities through the design of holistic solutions (e.g., Jagtap and Larsson 2019; Brugmann and Prahalad 2007). Cocreation between marginalised communities and partners such as NGOs, companies, and governments can generate greater value for each of them (e.g., Jagtap 2019a).

Social integration: Codesign with people living in resource-limited societies can potentially develop social embeddedness of outside-partners (e.g., Jagtap 2022a). This social embeddedness can further support codesign activities of resource-poor people.

3.1.1.2 Development of Resource-Limited Societies

The outcomes related to development of resource-limited societies are about empowerment of resource-poor people, their skill development, generation of income and livelihood opportunities, gender equity, and acceptance and use of solutions (e.g., Jagtap 2022a).

Empowerment of resource-poor people: A number of studies, carried out in a broad range of sectors such as energy, agriculture, craft, and healthcare, have identified several benefits of codesign for resource-constrained people. Codesign has the following beneficial outcomes for resource-poor people (e.g., Jagtap 2019a; Gonzalez et al. 2017).

- Codesign has the ability to empower resource-constrained people for working on design tasks.
- Codesign can train resource-poor people for further participatory tasks.
- Codesign can potentially develop and enhance design capability of resource-constrained people.
- Project ownership of resource-poor people can increase as they can potentially develop a feeling that their inputs and views are valuable in codesign activities.

Table 3.2 Outcomes of codesign activities for various stakeholders

Stakeholder	Outcomes of codesign activities
Resource-constrained individuals	• Cultivating a sense of ownership over projects • Improving design skills and capabilities • Potential opportunities for income generation • Empowerment
Designers and other partners	• Developing comprehensive understanding of the resource-constrained communities • Gaining deeper understanding of strengths and limitations of these communities • Gaining sound knowledge about life circumstances and requirements of people living in these communities • Combining knowledge, expertise, and available resources to develop solutions • Meeting the unfulfilled requirements of resource-limited individuals through collaborative design

Skill development: Codesign can also support people living in resource-constrained communities in developing some specific skills depending on the nature of the participatory project. For example, they can gain training in ICT (Hooli et al. 2016) or they can develop human interaction skills (Jagtap 2022a).

Income and livelihood generation: Codesigning solutions can support generation of income and livelihood opportunities for people living in resource-limited communities and can also develop and enhance their entrepreneurial abilities (Jagtap 2022a).

Gender equity: A study undertaken by Wood (2016) suggests that codesign activities can support female participants not only in addressing gender-related dynamics in projects but also in applying their skills for the benefit of their community.

Acceptance and use of solutions: Several studies have suggested that codesign activities can positively influence acceptance and use of designed solutions (e.g., Jagtap 2019a; Brubaker et al. 2017; Martins et al. 2018).

The above findings suggest that the outcomes of codesign activities are valuable not just for resource-constrained people but also for designers and other stakeholders. These outcomes for various stakeholders are summarised in Table 3.2.

3.1.2   How to Codesign in Low Resource Settings

The research in this field has offered several suggestions for codesigning with resource-limited individuals. While some suggestions focus on different types of resources and methods that can aid codesign activities with these people, others centre on factors that enable such codesign activities. In addition, the literature also advocates considering factors that can hinder codesign with resource-limited people since the knowledge on these hindering factors can support organisations to explore and implement ways to address such impeding factors. In the following paragraphs, these suggestions about resources and methods, enabling factors, and hindering factors are presented.

3.1.2.1 Resources and Methods Required for Codesign

Table 3.3 shows resources and methods required for codesigning with people living in resource-limited societies. A broad range of resources and methods can support this codesign, including codesign methods, contextual inputs, and organisational resources (Jagtap 2022a). These are discussed in the paragraphs that follow.

Codesign Methods

Codesign methods support a broad range of tasks, namely involving resource-constrained individuals in design projects and supporting sharing of information between them and other partners such as businesses, NGOs, and local governments (Jagtap 2021, 2022b). In addition to these activities, codesign methods play a vital role in supporting resource-constrained people in understanding problems, and in encouraging them to develop and assess ideas to tackle problems (Jagtap 2020). Codesign methods and resources also alleviate the power-inequality that can exist between marginalised individuals and organisations involved in designing solutions. Extant research has reported some codesign methods. They are as follows:

Graphical methods: Illiteracy, semi-literacy, and innumeracy rates are high in marginalised communities as people from these communities face a multitude of problems in completing their education (Prahalad 2004). As such, codesign methods using drawings, sketches, and pictures support organisations to communicate complex information to marginalised people and can support these people in understanding problems as well as possible solutions (Jagtap 2020). Such graphical codesign methods also help local people in sharing their views with organisations.

Storytelling strategies: In addition to codesign methods based on the use of pictures and drawings, narrative methods of sharing views and ideas are also effective in codesigning with marginalised individuals (Jagtap 2020). Narrative codesign methods

Table 3.3 Resources and methods to codesign in low resource settings (e.g., Jagtap 2022a)

Category	Resources and methods
Codesign methods	• Graphical methods • Storytelling strategies • Role-playing and design games
Contextual inputs	• Methods grounded in local strengths • Adjusting methods to local and project specificities
Organisational resources	• Financial and human resources

typically use stories and examples. They generally avoid discussions that are abstract in nature or lack easy-to-understand examples.

Role-playing and design games: Some studies have reported codesign methods that use strategies such as role-playing, prototyping, design games, and examples of using solutions (Jagtap 2022a; Dayaratne 2016; Jakobsson and Pekkala 2015).

Contextual Inputs

Inputs required to undertake codesign with resource-poor people are more effective when they are adapted to the target context. These contextual inputs are as follows:

Methods grounded in local strengths: In addition to codesign methods that use narrative ways, pictures, drawing, and design games, codesign activities can benefit from strengths of local communities (Jagtap 2022b). For instance, people living in marginalised communities help each other in numerous tasks. They also rely on their local networks for seeking information as well as for common community-level tasks (e.g., Nakata and Weidner 2012). Codesign activities can gain from such local strengths, for example, to spread awareness about new projects and to encourage involvement of local people in codesign tasks.

Adjusting methods to local and project specificities: There is diversity across marginalised communities. For example, needs, problems, and existing resources and capabilities in one community can be different from those in some other community. As such, it is crucial to consider this diversity in selecting or devising appropriate codesign methods for a given community and specific characteristics of a design project. In other words, adaptation of codesign methods to the project requirements and for satisfying needs and specificities of a target context is an essential aspect to involve marginalised people in design and development tasks (Dayaratne 2016; Jagtap 2022a; Arrivillaga et al. 2020). For example, adaptation of codesign methods to local conditions and project specificities has been reported by Dayaratne (2016) and Arrivillaga et al. (2020) in their affordable housing and healthcare projects, respectively.

Organisational Resources

In addition to contextual inputs and codesign methods, inputs related to organisational resources play a crucial role in undertaking codesign activities in this field. One such input is as follows:

Financial and human resources: To undertake activities in various codesign phases, organisations such as for-profit companies and NGOs ought to allocate necessary human and financial resources (Jagtap 2022a). For example, these organisations ought to assign someone the role of initiating, managing, and maintaining codesign activities. Furthermore, organisations ought to allocate resources to enhance their presence and embeddedness in local communities as it supports codesigning with local people. Organisations also need to devote time and resources to actively participate in community festivals because such participatory activities help in starting and managing codesign project.

The above discussion suggests that many different types of methods and resources can assist in involving marginalised people in design tasks (Jagtap 2022a). They can manifest in the form of graphical methods, storytelling strategies, design games, methods relying on local strengths, and financial and human resources. Knowledge about various codesign methods and resources can support the practice of codesign in this field. For example, this knowledge can assist practitioners in selecting and adapting codesign methods based on the requirements and circumstances of a target context and project. In the case of some contexts and projects, practitioners may not have resources for employing codesign methods that use pictographic media or prototypes. In this case, they can gain by building on social strengths of the community as these strengths are typically present in marginalised communities. Practitioners ought to provide some form of compensation to marginalised people when they employ social strengths in the community. They also ought to compensate for the time and resources marginalised people devote towards codesign tasks.

3.1.2.2 Enabling Factors

Table 3.4 presents the factors that support codesigning with resource-constrained people. Of the various supporting factors, two factors relate to the target context (integration in the local context, local language and dialects), three to the aspects of codesign methods (experience of codesign, gender-sensitive techniques, codesign mindset), two to organisational aspects (incentives and rewards, long-term perspective and patience), and three to partnership dimensions (trust, feedback and successful outcomes, rapport with local opinion leaders). The sections that follow present the details of the above-mentioned supporting factors.

Target Context

Two enabling factors—'integration in the local context' and 'local language and dialects'—are about the target context of resource-limited societies.

Table 3.4 Enablers in codesigning with people living in resource-constrained societies (e.g., Jagtap 2022a, 2022b)

Category	Enabling factors
Target context	• Integration in the local context • Local language and dialects
Codesign methods	• Experience of codesign • Gender-sensitive techniques • Codesign mindset
Organisational factors	• Incentives and rewards • Long-term perspective and patience
Partnership dimensions	• Trust • Feedback and successful outcomes • Rapport with local opinion leaders

Integration in the local context: Local embeddedness and integration in target communities can support codesign activities. Marginalised people perceive a locally embedded organisation as a part of their community. This not only helps the organisation in building and maintaining trusting relationships with the community, but also helps in gaining rich insights into needs of the community (Jagtap 2022a). Marginalised people willingly participate in cocreation activities in projects that focus on addressing their needs (e.g., Jagtap 2022b; Zanetell and Knuth 2004).

Local language and dialects: Previous studies have reported that employing local dialects and language supports codesign in this field (e.g., Jagtap 2022b; Bharti et al. 2014).

Codesign Methods

Three enabling factors—'experience of codesign', 'gender-sensitive techniques', and 'codesign mindset'—relate to the aspects of codesign methods.

Experience of codesign: Experience of involving marginalised people in design activities is a factor that enables codesign in this field. For example, organisations having codesign experience in marginalised societies gain knowledge about what works and what does not work in involving people from these societies in design projects (Jagtap 2022a). This supports these organisations in identifying and using effective codesign strategies. Likewise, codesign experience can also inspire marginalised people in contributing towards design projects (e.g., Bharti et al. 2014).

Gender-sensitive techniques: Employing gender-sensitive codesign processes and techniques has a beneficial impact on involvement of women from marginalised communities in design projects. This can be achieved through several ways. An empirical study, undertaken by Jagtap (2022b), suggests that when participatory activities are managed by female designers, women from the community can willingly participate in projects and share information and their perspectives. In addition to this technique, implementing rules of law that support rights of women to overcome extant stereotypes about gender-based views and inequalities can also encourage women to participate and contribute towards design activities.

Codesign mindset: Positive attitude towards working and cocreating with marginalised communities has a beneficial impact on codesign in this field (e.g., Jagtap 2022b). Flexible mindset and the recognition that marginalised people are experts in living in resource-constrained settings have beneficial impact on codesign in this field (e.g., Jagtap 2022b; Murcott 2007).

Organisational Factors

Two enabling factors—'incentives and rewards' and 'long-term perspective and patience'—relate to the organisational aspects.

Incentives and rewards: An empirical study in organisations that used codesign or participatory activities with marginalised people suggests that providing incentives to marginalised people encourages them to participate in projects (Jagtap 2022b).

Projects that deal with addressing pressing needs of marginalised communities can serve as incentives. Monetary compensation can also serve as an enabling factor in codesigning with marginalised people.

Long-term perspective and patience: One of the important aspects of codesign in this field is to implement understanding that marginalised people have many duties to perform than codesign activities (Wood 2016). It is therefore important to maintain a long-term view and work with the community with patience and at their speed.

Partnerships dimensions

Three enabling factors—'trust', 'feedback and successful outcomes', and 'rapport with local opinion leaders'—relate to the dimensions of partnerships.

Trust: Trust between marginalised people and organisations such as companies and NGOs positively affects codesign activities (e.g., Jagtap 2020; Bharti et al. 2014). For example, when marginalised people trust these organisations, they can share relevant information without reluctance and can willingly participate in design tasks (Jagtap 2020). While trusting relationships help in codesigning with marginalised people, such relationships between partners such as companies, governments, and NGOs also support cocreation (e.g., Letaifa and Reynoso 2015; Sarmiento et al. 2020). Trusting relationships with marginalised communities can be supported by:

- Maintaining transparency in the ways in which organisations perform their work (e.g., Brugmann and Prahalad 2007).
- Explicitly and clearly informing marginalised people project objectives (e.g., Jagtap 2022a).
- Disclosing all crucial and relevant information (e.g., Jagtap 2020).

Feedback and successful outcomes: Giving feedback to marginalised people on the importance of their contribution in completed participatory project activities can stimulate their desire to participate in projects that may be undertaken in future (e.g., Jagtap 2022b). Successfully and participatorily completed projects can also inspire marginalised people to engage in design tasks of current and future projects.

Rapport with local opinion leaders: In the context of marginalised communities, views and opinions of key people such as local leaders and well-recognised or respected people can influence the ways in which local decisions are made. It is therefore crucial to build and maintain rapport with such key people. This can enable organisations in promoting cocreation activities in the community (e.g., Bharti et al. 2014).

3.1.2.3 Codesign Barriers

Previous studies have reported a number of factors hindering codesign with people living in resource-constrained societies (see Table 3.5). The barriers can negatively influence organisations' access to these people, can limit their regular participation in design projects, or can restrict their contribution to design tasks. Of the various hindering factors, four factors relate to the target context (deficiencies in knowledge, gendered power differences, irregular and unpredictable participation, remoteness), one to the aspects of codesign methods (lack of common ground), three to organisational aspects (limited resources, limited time, missing organisational backing), one to partnership dimensions (power differences and hierarchies). These inhibiting factors are discussed in the sections that follow.

Target context

Four impeding factors—'deficiencies in knowledge', 'gendered power differences', 'irregular and unpredictable participation', and 'remoteness'—relate to the context of resource-limited societies.

Deficiencies in Knowledge: People living in marginalised communities commonly suffer from low literacy or illiteracy and from innumeracy due to weak educational infrastructure and lack of access to schooling (e.g., Jagtap et al. 2013). These problems as well as their unfamiliarity with design projects can negatively impact their contribution to participatory tasks (e.g., Bharti et al. 2014; Jagtap 2020).

Gendered power differences: There can be differences in power structures between men and women living in marginalised communities. These differences can have limiting effect on participatory activities. Codesigning with women can be difficult, and necessitates appropriate codesign methods and sensitivity (e.g., Shroff and Kam 2011). Prior studies have revealed difficulties experienced by women in engaging in codesign activities, for example, gendered power differences may not support women to speak and engage in cocreation (Wood 2016).

Irregular and unpredictable participation: Factors such as family responsibility, illness, urgent need to find livelihood opportunity, etc. contribute towards irregular and unpredictable participation of marginalised people in design projects (e.g., Wood 2016; Jagtap 2022a).

Table 3.5 Barriers in codesigning with people living in resource-constrained societies (e.g., Jagtap 2022a, 2022b)

Category	Barriers
Target context	• Deficiencies in knowledge • Gendered power differences • Irregular and unpredictable participation • Remoteness
Codesign methods	• Lack of common ground
Organisational factors	• Limited resources • Limited time • Missing organisational backing
Partnership dimensions	• Power differences and hierarchies

Remoteness: Prior studies have revealed difficulties in codesigning with marginalised people living in remote villages and locations (e.g., Jagtap 2022b).

Codesign Methods

One impeding factor—'lack of common ground'—is about aspects of codesign methods.

Lack of common ground: When there is mismatch between aims of a design project and needs of marginalised people, codesign activities cannot be effectively undertaken (e.g., Jagtap 2022b).

Organisational Factors

Three barriers—'limited resources', 'limited time', and 'missing organisational backing'—relate to organisational issues.

Limited resources: In order to undertake codesign activities with marginalised people, organisations need some essential resources. As such, the lack of essential resources hinders involvement of marginalised people in design activities (e.g., Sarmiento et al. 2020; Sushama et al. 2018; Jagtap 2022b). Absence of resources or limited resources can result into (Jagtap 2022a):

- Inconsistent and less effective involvement of marginalised people in design activities.
- Limited involvement of marginalised people, e.g., participation of less than requisite number of people.
- Restricted ability in employing appropriate codesign methods and techniques.

Limited time: Similar to negative impact of resource constraints, time constraints also have limiting effect on codesign in this field. For example, time constraints can result into (e.g., Jagtap 2022b):

- Limited exploration and generation of concepts in the design process.
- Limiting effect on discussion and resulting shared understanding during codesign activities with marginalised people.

Missing organisational backing: While support from organisations is crucial for undertaking codesign activities with resource-constrained people, some organisations may not provide necessary support, for example, by not providing priority to such codesign activities. Under such circumstances, designers may carry out participatory activities just for illustration (Jagtap 2022a). In a similar fashion, researchers working in academic institutes may also be constrained by administrative processes as well as by routine norms of knowledge (e.g., Wood 2016). This may not support researchers to undertake participatory activities to full extent, and they may stop these activities when they collect data essential to address their academic obligations.

Partnerships Dimensions

One barrier—'power differences and hierarchies'—relates to the issues associated with partnerships.

Power differences and hierarchies: Presence of hierarchies and power differences in communities and among those participating in codesign activities can negatively impact codesign and its outcomes. This is exemplified as follows:

- Both privileged and underprivileged people may participate in some projects. In the case of such projects, people who are perceived as powerful and educated can influence participatory activities and related dialogues in order to fulfil their own requirements (Wood 2016; Jagtap 2020).
- There can be differences between resources available in NGOs and businesses. This can create power differences between them, affecting contribution of NGOs in participatory activities (e.g., Nahi 2018).

Many different factors can negatively affect codesign with marginalised people (e.g., Jagtap 2022a and Jagtap 2022b). A broad range of constraints and deprivations in marginalised communities such as low literacy level and innumeracy can potentially be responsible for some of these barriers. On the other hand, some barriers can be related to aspects of codesign methods and processes such as mismatch between project aims and needs of marginalised people. Likewise, aspects of organisational issues such as limited resources and lack of support can also pose difficulties in codesigning with marginalised people. Similarly, collaboration-related aspects such as differences in power structures and presence of hierarchy can hinder codesign tasks.

Organisations can enhance codesign efforts by employing certain factors. Knowledge about the above-mentioned barriers and enablers can help organisations to use enablers to overcome barriers. For instance, they can use training programmes as a tool to overcome knowledge constraints in limited resources. By designing and implementing appropriate training programmes, they can support individuals to enhance their design knowledge and skills for actively contributing to design activities.

The tasks associated with codesign ought to be carried out in all phases of the design process (Jagtap 2021). This is illustrated in Fig. 3.1, together with some methods than can support these tasks.

Box 3.1 Codesign in Low Resource Settings—A Case Example

A Mexican company Cemex established Patrimonio Hoy in order to satisfy housing-related needs of low-income people (London 2008). Firstly, Patrimonio Hoy aimed at modifying its existing products and solutions for addressing the needs of low-income individuals, without using participatory design activities. Patrimonio Hoy's solution, devised without participatory design activities, was not successful (Jagtap 2019b). The design of the solution, lacking codesign with low-income people, comprised of its current product with some changes. Following this, Patrimonio Hoy involved low-income people in various design activities, supporting the organisation in gaining insights into constraints and problems faced by these people. The solution, devised with participatory design activities, consisted of several aspects tailored

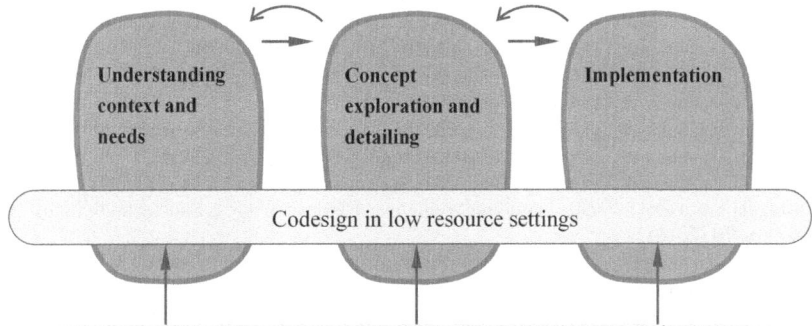

Codesign in low resource settings

How to co-design in low resource settings
- Codesign Methods (e.g., graphical methods, storytelling strategies, etc.)
- Methods grounded in local strengths
- Adjusting methods to local and project specificities
- Appropriate financial and human resources
- Taking into account barriers and enablers of co-design

Fig. 3.1 Design process and methods for codesigning in low resource settings

to the needs and conditions of low-income people. The solutions focused on construction of homes rather than selling cement. Additionally, it supported the people in gaining microloans. In the solution, the organisation also included do-it-yourself strategies that provided low-income people technical support for home construction.

References

M. Arrivillaga, P.C. Bermúdez, J.P. García-Cifuentes, J. Botero, Innovative prototypes for cervical cancer prevention in low-income primary care settings: A human-centered design approach. PLoS ONE **15**(8), e0238099 (2020)

S. Ben Letaifa, J. Reynoso, Toward a service ecosystem perspective at the base of the pyramid. J. Serv. Manag. **26**(5), 684–705 (2015)

K. Bharti, R. Agrawal, V. Sharma, What drives the customer of world's largest market to participate in value co-creation? Mark. Intell. Plan. **32**(4), 413–435 (2014)

E.R. Brubaker, C. Jensen, S. Silungwe, S.D. Sheppard, M. Yang, Codesign in Zambia-an examination of design outcomes. Int. Conf. Eng. Des. Vancouver (2017)

J. Brugmann, C.K. Prahalad, Cocreating business's new social compact. Harv. Bus. Rev. **85**(2), 80 (2007)

L. Corsini, S. Jagtap, J. Moultrie, Design with and by marginalized people in humanitarian makerspaces. Int. J. Des. (2022)

R. Dayaratne, Creating places through participatory design: Psychological techniques to understand people's conceptions. J. Housing Built Environ. **31**(4), 719–741 (2016)

M. De Silva, Z. Khan, T. Vorley, J. Zeng, Transcending the pyramid: Opportunity co-creation for social innovation. Ind. Mark. Manage. **89**, 471–486 (2020)

C.E.R. Gonzalez, P. Divigalpitiya, T. Sakai, The potential of participatory design to improve urban spaces in the slums of Caracas, Venezuela. Sustain. Dev. Plan IX **226**, 469 (2017)

L. Hooli, J.S. Jauhiainen, K. Lähde, Living labs and knowledge creation in developing countries: Living labs as a tool for socio-economic resilience in Tanzania. Afr. J. Sci. Technol. Innov. Dev. **8**(1), 61–70 (2016)

S. Jagtap, Design and poverty: A review of contexts, roles of poor people, and methods. Res. Eng. Design **30**(1), 41–62 (2019a)

S. Jagtap, Key guidelines for designing integrated solutions to support development of marginalised societies. J. Clean. Prod. **219**, 148–165 (2019b)

S. Jagtap, Barriers and enablers in codesigning with marginalised people, in *International Design Conference—Design 2020*. Dubrovnik, Croatia (2020)

S. Jagtap, Frugal-IDeM: An integrated methodology for designing frugal innovations in low-resource settings, in *Design for Tomorrow* (Springer, Singapore, 2021), pp. 41–51

S. Jagtap, A. Larsson, P. Kandachar, Design and development of products and services at the base of the pyramid: A review of issues and solutions. Int. J. Sustain. Soc. **5**(3), 207–231 (2013)

S. Jagtap, T. Larsson, Design and frugal innovations: Three roles of resource-constrained people, in *DS 92: Proceedings of the DESIGN 2018 15th International Design Conference* (2018), pp. 2657–2668

S. Jagtap, Codesign in resource-limited societies: Theoretical perspectives, inputs, outputs and influencing factors. Res. Eng. Design **33**, 191–211 (2022a)

S. Jagtap, Co-design with marginalised people: Designers' perceptions of barriers and enablers. CoDesign **18**(3), 279–302 (2022b)

S. Jagtap, T. Larsson, Resource-limited societies, integrated design solutions, and stakeholder input. She Ji: J. Des. Econ. Innov. **5**(4), 285–303 (2019)

M. Jakobsson, J. Pekkala, The design approach to developing renewable energy systems in BoP markets. Int J Des Manag Prof Pract **8**(3), 63–78 (2015)

T. London, Cemex's Patrimonio Hoy: at the tipping point. Case Study No. 1-428-606 (Global Lens, William Davidson Institute, University of Michigan, 2008)

R. Martins, J.A. Cherni, N. Videira, 2MBio, a novel tool to encourage creative participatory conceptual design of bioenergy systems—The case of wood fuel energy systems in south Mozambique. J. Clean. Prod. **172**, 3890–3906 (2018)

S. Murcott, Co-evolutionary design for development: Influences shaping engineering design and implementation in Nepal and the global village. J. Int. Dev. **19**(1), 123–144 (2007)

T. Nahi, Co-creation for sustainable development: The bounds of NGO contributions to inclusive business. Bus Strategy Dev **1**(2), 88–102 (2018)

C. Nakata, K. Weidner, Enhancing new product adoption at the base of the pyramid: A contextualized model. J. Prod. Innov. Manag. **29**(1), 21–32 (2012)

C.K. Prahalad, *The fortune at the bottom of the pyramid: Eradicating poverty through profits* (Wharton School Publishing, Upper Saddle River, 2004)

I. Sarmiento, G. Zuluaga, S. Paredes-Solís, A.M. Chomat, D. Loutfi, A. Cockcroft, N. Andersson, Bridging Western and Indigenous knowledge through intercultural dialogue: Lessons from participatory research in Mexico. BMJ Glob. Health **5**(9), e002488 (2020)

G. Shroff, M. Kam, Towards a design model for women's empowerment in the developing world, in *Proceedings of the SIGCHI Conference on Human Factors in Computing Systems* (ACM, 2011), pp. 2867–2876

P. Sushama, C. Ghergu, A. Meershoek, L.P. de Witte, O.C. van Schayck, A. Krumeich, Dark clouds in co-creation, and their silver linings: Practical challenges I faced in a participatory project in a resource-constrained community in India, and how I overcame (some of) them. Glob. Health Action **11**(1), 1421342 (2018)

L. Wood, Community development in higher education: How do academics ensure their community-based research makes a difference? Commun. Dev. J. **52**(4), 685–701 (2016)

B.A. Zanetell, B.A. Knuth, Participation rhetoric or community-based management reality? Influences on willingness to participate in a Venezuelan freshwater fishery. World Dev. **32**(5), 793–807 (2004)

Chapter 4
Design Strategies for Training, Awareness, and Income Generation

Abstract This chapter discusses three key guidelines to support the integrated process of designing, developing, and implementing inclusive solutions tailored to resource-constrained communities. Firstly, the chapter presents details of the guideline for designing context-specific training programmes to address knowledge and skill gaps of various stakeholders, including resource-constrained individuals and communities. Secondly, it discusses the guideline about designing contextualised awareness programmes to serve specific conditions and literacy levels of resource-constrained individuals, ensuring successful acceptance and continued usage of solutions. Lastly, the chapter presents the specifics of the guideline about the generation of income opportunities in alleviating several challenges faced by resource-constrained communities. For each of these three guidelines, the chapter includes a case example, while offering recommendation on how the guidelines can be implemented in a project.

4.1 Context-Sensitive Training Programmes

Appropriately designed training programmes are imperative to overcome marginalised individuals' and other actors' deficiencies in knowledge and skills required to perform a range of activities in the design process and life cycle stages of the solution (Jagtap 2019b).

4.1.1 Why Tailored Training Programmes Are Needed

There are several reasons for the necessity of tailored training programmes in the context of resource-constrained communities (Jagtap 2019b). These are as follows. The design of solutions for marginalised societies and for performing various life cycle activities such as implementation and maintenance of designed solutions demand a broad range of knowledge and skills. As such, these efforts require inputs from many stakeholders (e.g., Sousa-Zomer and Miguel 2016; Jagtap et al. 2013). These stakeholders possess unique set of knowledge and skills depending on where they are socially embedded and their professional education. For example, stakeholders such as local NGOs are socially embedded in marginalised contexts. As such, they are typically familiar and knowledgeable about the local context (e.g., Jagtap et al. 2013). In contrast, stakeholders such as professionally trained designers and companies are typically 'outsiders' and are not locally embedded in marginalised societies. However, they can possess knowledge and skills in areas such as technology, design, and needs and preferences of consumers in global or non-local markets (e.g., Jagtap et al. 2013). While these different stakeholders have their unique strengths, they may not possess all the necessary knowledge and skills required to design solutions, implement them, and perform other life cycle activities. These actors require training to improve their skills and fill the gaps in their abilities, which will enhance their knowledge and performance in specific areas.

Marginalised individuals often struggle with technology-related knowledge and skills, making it challenging for them to comprehend the installation and maintenance procedures of the proposed solutions (e.g., Jagtap 2019b; dos Santos et al. 2014). This lack of technical know-how can hinder their ability to benefit from the solutions using technology. To ensure that marginalised individuals can assist in essential functions like implementing, maintaining, or using solutions, training programmes may have to address the obstacles related to their illiteracy, innumeracy, and inadequate knowledge. NGOs and local governments may also need adequate training to develop and improve their abilities and expertise in essential fields, similar to other stakeholders. This can assist them in effectively carrying out their roles and responsibilities.

Numerous studies have highlighted the need of tailored training programmes for marginalised communities, and, if necessary, for other stakeholders in order to satisfy a variety of functions in the design and development process and in other activities in the lifecycle of solutions aimed at supporting these communities (see Table 4.1). Customised training programmes that consider the unique characteristics and circumstances of the target environment not only assist in meeting diverse needs but also aid in empowering disadvantaged individuals by improving their competencies and know-how, thus enabling them to develop, reinforce, and sustain their abilities over an extended period.

4.1.2   How to Design Context-Sensitive Training Programmes

The following methods and techniques can support the design of training programmes in the context of low resource settings (e.g., Jagtap 2019b).

Developing local trainers: In developing training programmes in the context of marginalised communities, some studies suggest that an effective strategy is to train a few individuals within the community in the initial phase (Jagtap 2019b). These trained individuals can then serve as valuable resource, sharing their knowledge and

Table 4.1 Studies supporting the need of training programmes in the context of low resource settings

Author(s)	Context of the study and support for training
Jagtap et al. (2013)	Examined a large number of design solutions targeted towards improving the conditions of marginalised communities in various developing countries. Findings reveal that the majority of the solutions focused on training programmes to address the knowledge and skill gaps of the stakeholders involved.
London et al. (2010)	Findings of this large-scale study show that providing technical support and training for marginalised producers is an essential part of overcoming various productivity and transactional constraints they face.
Devisscher and Mont (2008)	The significance of training in coffee processing systems in Bolivia has been identified.
Friebe et al. (2013)	Training is essential in the context of solar home systems in Africa and Asia.
dos Santos et al. (2009)	Importance and necessity of training for do-it-yourself (DIY) furniture. Context is related to low-income people from Brazil.
Bengo and Arena (2013)	Need of training in the context of energy systems in Nicaragua, Nigeria, India, and Colombia.
Lemair (2009)	Training required in the context of solutions aimed at rural electrification in Zambia.
Ramachandran et al. (2012)	Training is important for supporting handloom artisans in India.

skills with other people in the community (Devisscher and Mont 2008). The local trainers' proficiency in their native language and their understanding of acceptable socio-cultural customs and norms help to improve the effectiveness and acceptance of the training programmes they deliver (e.g., Whitney 2010). Their unique perspectives and insights contribute to a more successful implementation of the programmes.

Incorporating living conditions and characteristics of trainees: In developing training programmes, it is essential to consider the living conditions of trainees to ensure that the programmes are effective (Jagtap 2019b). This includes taking into account a wide range of factors such as age, gender, educational background, and other demographic variables (e.g., Jagtap et al. 2013; Devisscher and Mont 2008). Disregarding these variables can result in programmes that do not meet the specific needs and circumstances of the trainees. One crucial factor to consider is the educational background of the trainees. Training programmes for individuals with lower levels of education may benefit from using more practical and hands-on training. Likewise, training programmes ought to consider specific needs and requirements of trainees from various age groups. Gender is another important variable to consider (Jagtap 2019a). Designing training programmes that are sensitive to preferences of men and women can ensure that all trainees benefit from the programme.

Concrete, local, and social: Training programmes that are designed for marginalised individuals require specific considerations in order to be effective (Jagtap 2019b). According to Viswanathan (2010), these programmes need to be

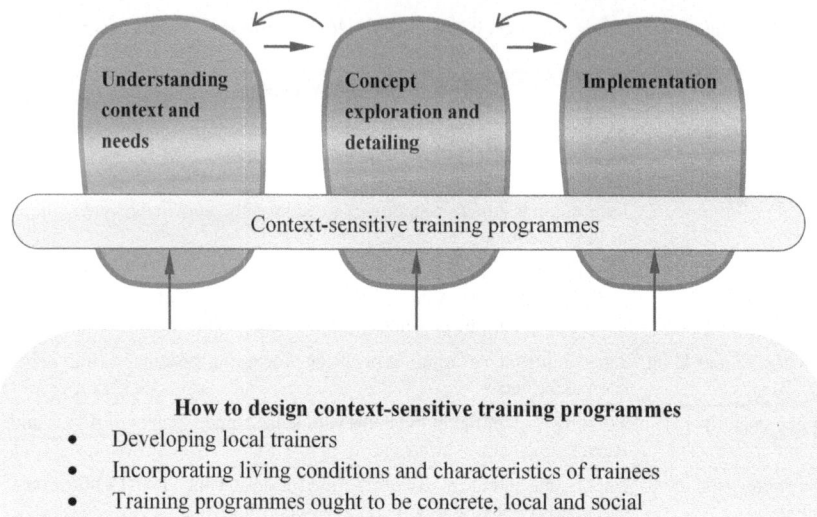

Fig. 4.1 Design process and methods for developing context-sensitive training programmes

concrete, local, and social. One important aspect is that training programmes for marginalised people need to be concrete, which involves adapting to the education and communication styles to accommodate their low literacy levels. Additionally, the programme should be tailored to the local context by utilising the local language and cultural nuances. Finally, a social aspect should also be incorporated into the programme by leveraging the socio-cultural strengths of the community. This approach can help individuals feel more connected and engaged in the learning process, leading to improved outcomes.

The tasks associated with the design and implementation of context-sensitive training programmes ought to be considered in all phases of the design process (Jagtap 2021). This is illustrated in Fig. 4.1, together with how some techniques and methods can support these tasks.

Box 4.1 Context-Sensitive Training Programmes—A Case Example

Bengo and Arena's (2013) study focuses on the innovative solution designed by DESI Power, a for-profit company in India, to address the challenges of providing electricity in rural areas. DESI Power's approach relies on the use of agro-residues and renewable energy to generate electricity. The company tailors its solutions to the specific needs of target rural areas by combining various renewable energy alternatives. DESI Power provides technical solutions based on biogas plants, renewable energy, and micro-generators to micro-enterprises requiring access to their energy solution. DESI Power has developed a specialised training programme aimed at bridging gaps in knowledge and skills within the value chain. This programme has been designed to meet specific requirements and needs of the recipients, ensuring that they receive tailored and effective training. DESI Power has created a training programme called 'DESI Power Mantra', which focuses on empowering local communities to develop skills and knowledge necessary to operate and maintain energy-generation plants. Through this programme, participants can build upon their existing capabilities and improve their capacity to manage plant operations. The company provides training to a local cooperative, enabling it to handle all aspects related to the activities associated with plant's installation and maintenance. As a part of its commitment to training and capacity building, the company sets aside approximately 10% of the total cost of each installation. This allocation is specifically for training purposes. DESI Power offers affordable solutions that address the energy needs of rural communities (Jagtap 2019b). These solutions are designed to be accessible and cost-effective, making them a viable option for people living in remote and underserved areas. Figure 4.2 offers an illustration of the above case example.

4.2 Designing Tailored Awareness Plans

Contextualising the design of awareness programmes to the life circumstances and literacy level of marginalised individuals and the broader context in which they live can enable successful dissemination and acceptance of designed solutions (Jagtap 2019b).

4.2.1 Why Context-Sensitive Awareness Plans Are Needed

Marginalised people face many different problems. One of these problems is about high illiteracy and knowledge deprivation (e.g., Ahmed et al. 2010; Vachani and Smith 2008; Nakata and Weidner 2012; Jagtap 2022). They often find themselves without adequate access to educational resources, which can result in a lack of knowledge and skills (Jagtap 2019b). This lack of education can lead to a number of other issues such as hunger and poor health, further exacerbating their cognitive difficulties. These difficulties can manifest in the form of memory retrieval problems and difficulty recalling available product options, among other things (e.g., Jagtap

Fig. 4.2 Illustration of the
DESI power case example

et al. 2013). They also face difficulties in reading information on product labels and selecting the most suitable option to fulfil a specific need (e.g., Viswanathan et al. 2005). In such cases, individuals may struggle to make informed decisions.

In addition to the above difficulties, people living in low resource settings may lack access to traditional promotion and awareness programmes (Jagtap 2019b). As a result, they cannot access and use knowledge that can help them in enhancing their performance in entrepreneurial activities or in promoting their well-being by adopting some essential healthcare practices. For example, they may not have means to access information about contemporary, affordable, and effective practices to increase farm yield or about benefits of adopting solutions that are environmentally sustainable (e.g., Prahalad 2004; UNDP 2008). In order to gain access and use a solution, it is essential to have knowledge about its availability and possible benefits.

The development of effective solutions to address the needs of marginalised populations requires a comprehensive understanding of their unique life circumstances, as well as their cognitive limitations resulting from illiteracy, innumeracy, and other cognitive barriers (Jagtap 2019b). Accordingly, it is essential to implement awareness programmes that are tailored to the specific needs of these individuals, building on their pictographic and concrete thinking ways. This can ensure that the designed

Table 4.2 Studies supporting the need of awareness programmes in the context of low resource settings

Study	Context of the study and indicative findings
Sousa-Zomer and Cauchick-Miguel (2016)	• Analysis of solutions in water and transportation sectors in Brazil • Designing suitable awareness programmes for marginalised communities has beneficial impact on sustained acceptance and usage of solutions
Jagtap et al. (2013)	• Analysis of many cases from several developing countries • Findings support a crucial requirement of designing programmes for raising awareness of relevant information in marginalised communities
Bengo and Arena (2013)	• Examination of energy systems from Nicaragua and India • Awareness programmes that take into account specific characteristics of the target context can support marginalised communities in accepting and using designed solutions
Ramachandran et al. (2012)	• Examination of a solution aimed at supporting handloom artisans in India • Awareness programmes for marginalised communities are important components of designed solution
Jagtap and Kandachar (2010)	• Analysis of a case from healthcare sector in Mali and from agriculture sector in Mexico • Contextualised awareness programmes have a beneficial impact on spread of solutions in marginalised communities • The awareness programmes play a key role in enhancing acceptance of design solutions

solutions are both relevant and accessible to them (e.g., UNDP 2008; Jagtap et al. 2013). May studies undertaken in a variety of contexts have supported the profound need of designing suitable awareness programmes for marginalised individuals and communities (see Table 4.2).

4.2.2 How to Create Context-Sensitive Awareness Plans

The following methods and techniques can support the design of awareness plans in the context of low resource settings (e.g., Jagtap 2019b).

Thinking styles: Individuals who experience marginalisation may have a greater tendency towards concrete and pictographic thinking (e.g., Jagtap and Larsson 2013). As such, awareness programmes designed for marginalised people ought to take into account these thinking styles. This can ensure that the information conveyed is more easily understood and retained by marginalised people who may have difficulty in comprehending abstract or complex concepts. Therefore, by taking into account their cognitive styles, awareness programmes can be more effective.

Demonstrations: Previous research suggests that an effective way to design awareness programmes for marginalised communities is to demonstrate the potential advantages of a solution and its functioning (Jagtap 2019b). This approach was employed by Godrej and Boyce in their initiatives to raise awareness among rural Indians about the benefits of the refrigeration product developed by the company (Whitney 2010). By implementing demonstration-based techniques, they were able to convey the product's potential benefits effectively.

Social methods: Whitney's (2010) study revealed that promoting designed solutions within a community can be effectively accomplished through social interactions and conversations. For instance, neighbours explaining the benefits of a product to other members of the community can lead to increased promotion of the solution. As such, social methods play a crucial role in the success of awareness programmes designed for marginalised communities (Jagtap 2019b).

Road shows, word-of-mouth strategies, and other methods: In designing awareness programmes, it is crucial to incorporate the use of the local language and take advantage of existing social interactions (e.g., Jagtap 2019a, 2019b). Methods such as road shows, word-of-mouth strategies, partnerships with opinion leaders, and direct discussion with marginalised individuals also support the development of awareness programmes. Employing these methods can enhance the effectiveness of developed awareness programmes. Figure 4.3 illustrates the above-mentioned methods for developing tailored awareness programmes in the context of low resource settings.

The tasks associated with the design and implementation of context-sensitive awareness programmes ought to be considered in all phases of the design process

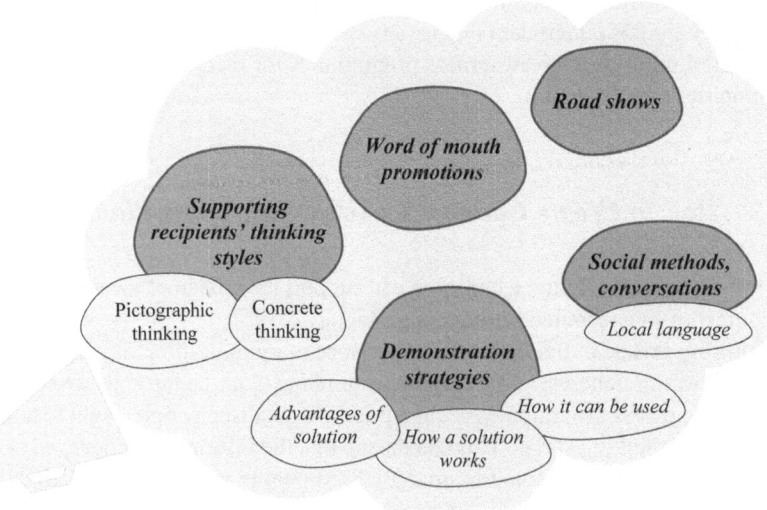

Fig. 4.3 Methods and techniques for developing tailored awareness programmes for low resource settings

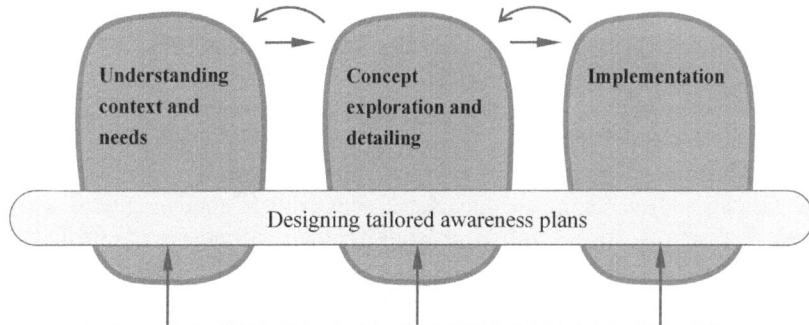

Fig. 4.4 Design process and techniques for developing context-sensitive awareness programmes

(Jagtap 2021). This is illustrated in Fig. 4.4, together with how some techniques and methods can support these tasks.

Box 4.2 Tailored Awareness Plans for Low Resource Settings—A Case Example

Amanco, a profit-driven enterprise, undertook the task of creating irrigation systems for impoverished farmers from the La Testaruda community in Mexico. This endeavour was achieved through the aid of various partners, such as the non-governmental organisation 'Sustainable Farmers Network' (RASA), with whom Amanco worked to design, develop, and ultimately deploy the irrigation systems (UNDP 2008; Jagtap and Kandachar 2010). In order to address the unique needs and circumstances of the La Testaruda community, Amanco designed and implemented awareness programmes tailored to their specific requirements. These awareness programmes included a number of strategies such as meetings, exhibitions, and word-of-mouth promotions, aimed at spreading the message and reaching many recipients. Exhibitions were designed with the intention of demonstrating and explaining workings and advantages of irrigation systems. These exhibitions were erected on plots provided by a few farmers, serving as a platform for educational demonstrations and explanations. In the process of designing promotion programmes, an important consideration was to avoid the direct selling of irrigation systems to low-income farmers. Instead, the focus was on providing support to these

farmers in recognising and understanding their day-to-day problems, ineffi-
ciencies in farming, and the potential benefits that irrigation systems could
offer to overcome these issues. The objective was to develop a more nuanced
understanding of the challenges faced by these farmers and tailor the promotion programmes accordingly. These programmes were designed to empower
the farmers and encourage them to take an active role in improving their agricultural practices. RASA utilised existing farming cooperatives to facilitate
their awareness and promotion programmes. When such cooperatives were
lacking, RASA actively encouraged their development. By partnering with
these cooperatives, RASA was able to reach out to low-income farmers and
provide them with the support they needed. Due to the implementation of
irrigation systems, there was a noticeable increase in agricultural production,
resulting in a threefold increase in the earnings of farmers. These proficient
systems not only helped decrease water usage, land erosion, and time required
for irrigation but also supported farmers to allocate more time towards other
farming responsibilities (Jagtap 2019b). Figure 4.5 illustrates the above case
example.

Fig. 4.5 Illustration of the
Amanco's awareness
programme

4.3 Integrating Income Opportunities

Increasing income of marginalised people is an effective way to alleviate their problems. Design solutions that create income generation opportunities are preferred by marginalised people and have greater impact on their lives (Jagtap 2019b).

4.3.1 Why to Integrate Income Opportunities in the Solution

There are several reasons why there is a crucial need for integrating income opportunities in solutions aimed at supporting betterment of people living in low resource settings (Jagtap 2019b). These reasons are elaborated as follows.

Low and unpredictable income—a tenacious deprivation: One of the most significant and enduring challenges faced by marginalised communities is the difficulty of getting a stable and reliable income (e.g., Chakravarti 2006). For these individuals, a significant portion of their earnings is often dedicated to meeting essential needs such as food and shelter. This means that they have limited financial resources to invest in other areas, such as education or skill development. In contrast to non-marginalised individuals, who typically have greater financial security and resources at their disposal, the income of marginalised communities is often characterised by unpredictability and volatility. Many of these individuals may struggle with persistent underemployment or unemployment (e.g., Nakata and Weidner 2012). Moreover, the stress of dealing with low and unpredictable income can have significant psychological and emotional impacts, leading to anxiety.

Importance of raising income: In resource-limited settings, income is considered a crucial dimension of scarcity (Jagtap 2019b). An individual's level of income can impact his or her ability to meet basic needs and access to resources. As a result, income is often used as a key metric to quantify the challenges faced by marginalised individuals. Karnani (2011) argues that raising income of marginalised people is an effective alternative to alleviate their difficulties. This is because higher income levels can enable people in gaining access to healthcare, education, and adequate housing. It can also enhance their autonomy and ability to make life choices. The importance of income is also recognised by the United Nations (UN 2015) through the inclusion of the target 'full and productive employment and decent work for all' in the sustainable development goals (SDGs). This target aims to promote access to decent work and income-earning opportunities for all individuals. Many studies have revealed that solutions providing income generation and livelihood opportunities are valued by marginalised people (see Table 4.3).

Table 4.3 Studies supporting integration of income opportunities in the solutions developed for low resource settings

Study	Context of the study and findings
López et al. (2017)	• Conducted a study in the field of renewable energy • Examined important parameters in designing solutions for marginalised communities • Examined marginalised people's preferences for attributes of solutions • Findings reveal that income generation opportunities offered by a solution are valued by marginalised people
Thomas (2006)	• Analysed designs aimed to satisfy needs of low-income people • Examined a product aimed at reducing physical burden of washerwomen • Findings reveal that economic gains afforded by the product were given higher preference by the washerwomen in evaluating the product
Ness and Xing (2010)	• Analysed several design solutions that were deemed as successful • Marginalised people value design solutions that help them to enhance their income directly or increase their economic benefits indirectly
Jagtap and Kandachar (2010)	• Examined irrigation systems designed by an organisation called Amanco • The irrigations systems aimed to support productivity and income of low-income farmers • Income opportunities in the solution appeared to be an important factor in motivating farmers to engage in implementation of the irrigation systems • Income opportunities can potentially inspire marginalised individuals to invest in the solutions and increase their willingness to pay

4.3.2 How to Integrate Income Opportunities

Employing marginalised people: In designing solutions for marginalised communities, it is advisable to generate livelihood prospects by employing marginalised individuals in various stages of the solution's life cycle, such as production, promotion, distribution, installation, and maintenance (e.g., Bacchetti 2017; Bengo and Arena 2013; Jagtap 2019b). In creating these employment opportunities, it is crucial to take into account literacy and skill levels of these individuals. By providing such opportunities, the designed solution can not only address the problem at hand but can also promote economic empowerment of marginalised communities.

Marginalised people as entrepreneurs: Another recommendation in the literature is about focusing on marginalised people as entrepreneurs engaged in production of some types of products and services (e.g., framers). When marginalised people are considered as producers, their income can be raised by supporting their related activities of producing products and services as well as by alleviating various productivity and transactional constraints they face (e.g., Webb et al. 2015).

Supporting entrepreneurial activities: Another suggestion in the literature is to support entrepreneurial activities of marginalised people, for example, by providing them access to electricity or lighting solutions (e.g., Lemaire 2009; Jagtap 2019b). It is widely recognised that people living in marginalised communities typically face numerous problems in their entrepreneurial activities. Lack of access to electricity is one such barrier. Therefore, providing them access to electricity or lighting

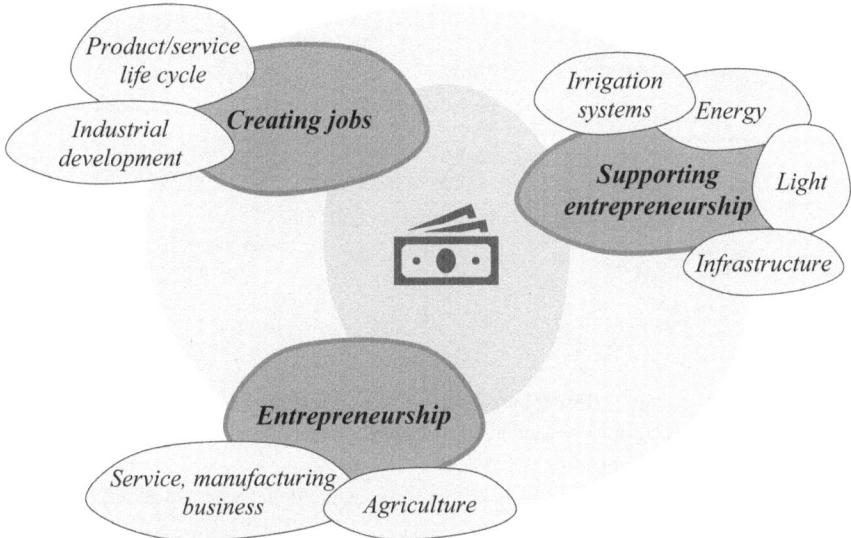

Fig. 4.6 Some methods for integrating income opportunities in solutions developed for low resource settings

solutions can alleviate this barrier and create new opportunities for income generation. Figure 4.6 illustrates the above-mentioned methods for integrating income opportunities in solutions developed for low resource settings.

The tasks associated with the integration of income opportunities in the solution ought to be considered in all phases of the design process (Jagtap 2021). This is illustrated in Fig. 4.7, together with how some methods can support these tasks.

Box 4.3 Integrating Income Opportunities—A Case Example

Chowdhury (2006) documented a project named 'Women as Solar Power Entrepreneurs'. The project was conducted by The Energy and Resources Institute (TERI). The project was targeted towards aiding the rural population residing in and around the Sundarbans region of West Bengal, India. One of the objectives of the project was to empower women by training them to become solar power entrepreneurs, thus promoting sustainable energy practices in the region. Local women were given training as a part of the project to assemble various products including solar lanterns, DC mobile chargers, home-lighting systems, and LED lamps. This empowered women by equipping them with the skills to assemble a diverse range of products. The project aimed to create job opportunities and promote self-reliance among women in the community. Moreover, the women received training in product installation, maintenance, and marketing, which also developed income opportunities. This not only enabled them to improve their financial gains but also facilitated the

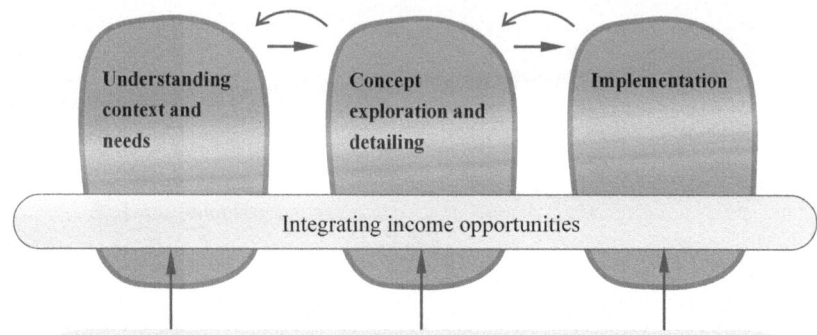

Fig. 4.7 Design process and methods for integrating income opportunities in the solution for low resource settings

dissemination of solar lanterns to the local community. Solar lanterns proved to be a valuable asset for farmers involved in betel leaf cultivation, a significant source of income in the region. The traditional use of kerosene lamps for leaf plucking at night had a detrimental effect on the quality of the produce, imparting an unpleasant kerosene odour. The introduction of solar lanterns as a substitute for kerosene lamps resulted in improved quality of betel leaves. The innovative solution devised by TERI not only resolved the issue of odour but also supported income opportunities for the farmers in the region (Jagtap 2019b).

References

A. Ahmed, L. Kabir, H. Yasuura, An information platform for low-literate villagers, in *2010 24th IEEE International Conference on Advanced Information Networking and Applications* (IEEE, 2010), pp. 1271–1277

V. Akula, Business basics at the base of the pyramid. Harv. Bus. Rev. Bus. Rev. **86**(6), 53 (2008)

E. Bacchetti, A design approach with method and tools to support SMEs in designing and implementing distributed renewable energy (DRE) solutions based on sustainable product-service system (S. PSS). Procedia CIRP **64**, 229–234 (2017)

I. Bengo, M. Arena, Integrating the social dimension into new business models for energy access, in *Renewable Energy for Unleashing Sustainable Development* (Springer, Cham, 2013), pp. 203–217

D. Chakravarti, Voices unheard: The psychology of consumption in poverty and development. J. Consum. Psychol.consum. Psychol. **16**(4), 363–376 (2006)

F. Chowdhury, *Sun Power, Women Power*, Boloji. http://www.boloji.com/wfs5/wfs625.htm/. Accessed December 2017 (2006)

T. Devisscher, O. Mont, An analysis of a product service system in Bolivia: Coffee in Yungas. Int. J. Innov. Sustain. Dev. **3**(3–4), 262–284 (2008)

A. Dos Santos, A. Krämer, C. Vezzoli, Design brings innovation to the base of the pyramid. Des. Manag. Rev. **20**(2), 78–85 (2009)

A. Dos Santos, C.P. Sampaio, J.S. Giacomini da Silva, J. Costa, Assessing the use of product-service systems as a strategy to foster sustainability in an emerging context. Product Manag. Dev. **12**(2), 99–113 (2014)

M. Esposito, A. Kapoor, S. Goyal, Enabling healthcare services for the rural and semi-urban segments in India: When shared value meets the bottom of the pyramid. Corporate Governance: Int. J. Bus. Soc. **12**(4), 514–533 (2012)

C.A. Friebe, P. von Flotow, F.A. Täube, Exploring the link between products and services in low-income markets: Evidence from solar home systems. Energy Policy **52**, 760–769 (2013)

S. Jagtap, Design and poverty: A review of contexts, roles of poor people, and methods. Res. Eng. Design **30**, 41–62 (2019a)

S. Jagtap, Key guidelines for designing integrated solutions to support development of marginalised societies. J. Clean. Prod. **219**, 148–165 (2019b)

S. Jagtap, Co-design with marginalised people: Designers' perceptions of barriers and enablers. CoDesign **18**(3), 279–302 (2022)

S. Jagtap, P. Kandachar, Representing interventions from the base of the pyramid. J. Sustain. Dev. **3**(4), 58–73 (2010)

S. Jagtap, A. Larsson, Design of product service systems at the base of the pyramid, in *ICoRD'13. Lecture Notes in Mechanical Engineering* ed. by A. Chakrabarti, R.V. Prakash (Springer India, 2013) pp. 581–92

S. Jagtap, A. Larsson et al., Design and development of products and services at the Base of the Pyramid: A review of issues and solutions. Int. J. Sustain. Soc. **5**(3), 207–231 (2013)

S. Jagtap, Frugal-IDeM: An integrated methodology for designing frugal innovations in low-resource settings, in *Design for Tomorrow—Volume 2: Proceedings of ICoRD 2021* (Springer Singapore, 2021), pp. 41–51

A. Karnani, *Fighting poverty together: Rethinking strategies for business, governments, and civil society to reduce poverty* (Palgrave Macmillan, New York, NY, 2011)

X. Lemaire, Fee-for-service companies for rural electrification with photovoltaic systems: The case of Zambia. Energy Sustain. Dev. **13**(1), 18–23 (2009)

T. London, R. Anupindi, S. Sheth, Creating mutual value: Lessons learned from ventures serving base of the pyramid producers. J. Bus. Res. **63**(6), 582–594 (2010)

A.M. López, F. Musonda, T. Sakao, N. Kebir, Lessons learnt from designing PSS for base of Pyramid. Procedia CIRP **61**, 623–628 (2017)

C. Nakata, K. Weidner, Enhancing new product adoption at the base of the pyramid: A contextualized model. J. Prod. Innov. Manag.innov. Manag. **29**(1), 21–32 (2012)

D. Ness, K. Xing, A theoretical construct of a product-service systems model for low social-economic status communities, in *INFORMS Service Science Conference*, Taipei (2010)

C.K. Prahalad, The fortune at the bottom of the pyramid: Eradicating poverty through profits (Wharton School Publishing, Upper Saddle River, NJ, 2004)

J. Ramachandran, A. Pant, S.K. Pani, Building the BoP producer ecosystem: The evolving engagement of Fabindia with Indian handloom artisans. J. Prod. Innov. Manag.innov. Manag. **29**(1), 33–51 (2012)

T.T. Sousa-Zomer, P.A.C. Miguel, Exploring the critical factors for sustainable product-service systems implementation and diffusion in developing countries: An analysis of two PSS cases in Brazil. Procedia CIRP **47**, 454–459 (2016)

A. Thomas, Design, poverty, and sustainable development. Des. Issues **22**(4), 54–65 (2006)

UN, *Transforming Our World: The 2030 Agenda for Sustainable Development* (2015)

UNDP (2008). Creating Value for All: Strategies for Doing Business with the Poor. United Nations Development Programme.

S. Vachani, N.C. Smith, Socially responsible distribution: Distribution strategies for reaching the bottom of the pyramid. Calif. Manage. Rev. **50**(2), 52–84 (2008)

M. Viswanathan, *A Micro-Level Approach to Understanding BoP Markets. Next Generation Business Strategies for the Base of the Pyramid: New Approaches for Building Mutual Value* (FT Press, Upper Saddle River, 2010)

M. Viswanathan, J.A. Rosa, J.E. Harris, Decision making and coping of functionally illiterate consumers and some implications for marketing management. J. Mark. **69**(1), 15–31 (2005)

J.W. Webb, C.G. Pryor, F.W. Kellermanns, Household enterprise in base-of-the-pyramid markets: The influence of institutions and family embeddedness. Africa J. Manag. **1**(2), 115–136 (2015)

P. Whitney, *Reframing Design for the Base of the Pyramid. Next Generation Business Strategies for the Base of the Pyramid: New Approaches for Building Mutual Value* (Upper Saddle River, New Jersey, pp. 165–192, 2010)

Chapter 5
Lifecycle Engineering and Intersectoral Collaboration

Abstract This chapter explores the following three key guidelines for designing, developing, and implementing integrated solutions for supporting social and human development of resource-limited societies. First, it highlights the critical guideline of identifying and implementing life cycle requirements in various phases of the design and development process. Secondly, it provides details of the guideline about collaborative design, highlighting the crucial role of partnerships between NGOs, local governments, and for-profit entities in addressing complex constraints and facilitating design of effective solutions. Finally, the chapter discusses the guideline about the importance of adapting project management to the specificities of local contexts, emphasising the crucial role of avoiding assumptions and biases in the process of designing appropriate solutions for resource-constrained societies. For each of these three guidelines, the chapter provides an illustrative example, while offering recommendations on how the guidelines can be implemented.

5.1 Enforcing Life Cycle Needs

Identifying and implementing life cycle requirements are essential tasks for continued use of the solution by marginalised communities (Jagtap 2019b).

5.1.1 Why to Implement Life Cycle Needs

There are several reasons for identification and implementation of life cycle needs in developing solutions for low resource settings (Jagtap 2019b). These are as follows.

Difficulties in undertaking life cycle activities: In marginalised societies, there are various constraints and deprivations that impede important activities in the life cycle of solutions. These constraints can have a negative impact on life cycle activities of a solution such as repair, maintenance, refurbishment, and recycling (UNDP 2008; Jagtap 2019b). Restoring a failed product to operational condition can be a

© The Author(s), under exclusive license to Springer Nature Switzerland AG 2024   63
S. Jagtap, *Design and Engineering for Low Resource Settings*,
SpringerBriefs in Applied Sciences and Technology,
https://doi.org/10.1007/978-3-031-66156-3_5

daunting task due to various factors such as inadequate management processes, weak infrastructure, and scarce availability of consumables and spare parts. Furthermore, the shortage of skilled workers to repair the product can exacerbate these challenges. Unfulfilled life cycle activities have a detrimental impact on sustained and long-term adoption and usage of designed solutions (Jagtap 2019a).

Failed and deteriorated products: There are many examples of products, covering several sectors such as agriculture and healthcare, that cannot serve the marginalised communities due to their failed or deteriorated condition (e.g., Jagtap and Kandachar 2010; Aranda-Jan et al. 2016; Jagtap 2019b). Similarly, numerous examples exist of distributed renewable energy (DRE) systems that have failed or become dysfunctional due to lack of requisite life cycle activities (e.g., Emili et al. 2016). The failure or dysfunction of DRE systems is not uncommon and can be attributed to two primary reasons: lack of corrective maintenance following a system failure, or lack of preventive maintenance to avoid breakdowns and extend system life. It is important for those responsible for DRE systems to prioritise both corrective and preventive maintenance to ensure optimal performance and longevity of these important energy systems.

Necessity of deliberate efforts to implement life cycle activities: In designing solutions aimed at supporting social and human development of marginalised communities, it is essential to take into account entire life cycle of the solutions (e.g., Ramani et al. 2012; Jagtap 2019b). This means taking into account requirements associated with life cycle activities such as repair, maintenance, refurbishment, and recycling in order to ensure that these solutions remain effective over time. Additionally, designers must consider the unique challenges faced by marginalised communities when implementing these life cycle issues, including limited resources. By identifying and incorporating life cycle requirements, designers can create more sustainable and impactful solutions for marginalised societies.

Realising life cycle requirements for sustained use of solutions: Based on their examination of solutions from several sectors such as water, transportation, and energy, several studies have established a critical need of realising requirements in the entire life cycle of a solution. One example is a bike-sharing solution in Brazil (Sousa-Zomer and Cauchick-Miguel 2017). The findings, based on the analysis of this example, reveal that addressing life cycle issues and implementing activities associated with preventive and corrective maintenance were crucial for achieving intended sustainable impact on the target context. Likewise, Jagtap and Kandachar (2010) propose that in designing solutions for underprivileged communities, a comprehensive approach that identifies and implements life cycle issues is crucial, as seen in their examination of irrigation systems designed for low-income farmers in Mexico. In a similar fashion, many case studies undertaken in energy sector reinforce the crucial need of recognising and realising pertinent life cycle requirements in order to support long-term usage of solutions by marginalised individuals (Bengo and Arena 2013; Bacchetti 2017). The above studies highlight the necessity of considering and implementing a wide range of life cycle requirements (e.g., Jagtap and Kandachar 2010; Aranda-Jan et al. 2016). Some studies suggest that incorporating life cycle issues can provide the following additional benefits.

- It increases life of a product, offering additional environmental and economic gains (Jagtap et al. 2013; Bacchetti 2017; Jagtap 2019b).
- It reduces the burden of manufacturing new products, reducing consumption of related material resources (Jagtap et al. 2013; Bacchetti 2017).
- It can result into affordable products through life cycle activities such as reuse, refurbishment, and remanufacturing (Jagtap 2019b).
- The implementation of life cycle activities also supports creation of employment opportunities for local marginalised people (Kamigaki et al. 2017).

5.1.2 How to Implement Life Cycle Needs

Table 5.1 presents three categories of alternatives that can support organisations in identifying and implementing life cycle needs in the process of designing and developing solutions for low resource settings. These categories are associated with: (1) identification of life cycle requirements, (2) using appropriate business models, and (3) design of appropriate products. For each of these categories, the table presents recommendations drawn from pertinent studies. Figure 5.1 illustrates the above information.

The tasks associated with identification and implementation of life cycle requirements ought to be carried out in all phases of the design process (Jagtap 2021). This is illustrated in Fig. 5.2, together with some methods than can support these tasks.

Table 5.1 Recommendations for identification and implementation of life cycle needs in the context of low resource settings

Category	Sources	Recommendations
Life cycle requirements	Aranda-Jan et al. (2016), Jagtap et al. (2013)	• It is crucial to consider requirements about life cycle issues early in the design process • Taxonomies of life cycle requirements can assist designers in considering related issues
Business models	Lemaire (2009), Kamigaki et al. (2017), Ness and Xing (2010)	• Life cycle issues can be addressed using appropriate business models • Business models in which solution provider is responsible for all the relevant life cycle activities such as installation of equipment, repair, maintenance, remanufacturing, disposal, etc., can support implementation of life cycle activities
Product design	Aranda-Jan et al. (2016), Jagtap et al. (2013)	• In designing products, consider and address aspects about robustness, reliability, and durability • Products need to be designed by taking into account dusty, hot, and other relevant conditions in marginalised societies • Products need to be easy to repair and maintain, within the constraints such as available maintenance infrastructure and skills of marginalised individuals • Assess products by testing them in actual setting. This can reveal issues (if any) regarding various life cycle phases

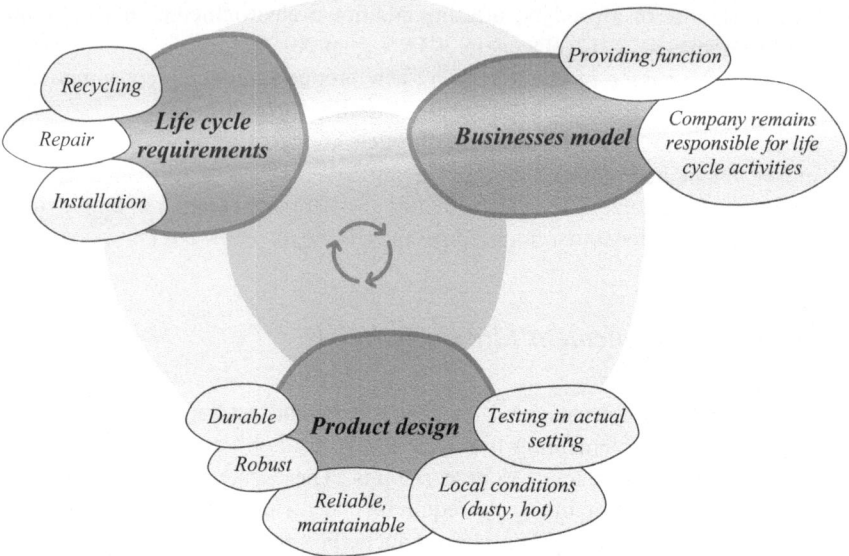

Fig. 5.1 Illustration of a few ways in which life cycle needs can be identified and implemented in the context of low resource settings

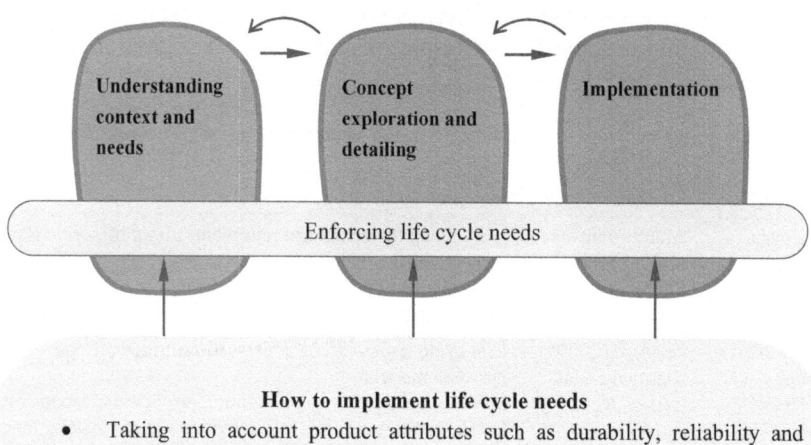

Fig. 5.2 Design process and methods for identification and implementation of life cycle requirements

Box 5.1 Enforcing Life Cycle Needs—A Case Example

A study, conducted by Sousa-Zomer and Cauchick-Miguel (2017), examined a water purification system that was implemented in Brazil. The system involves the installation of a water filtering equipment at designated sites, which can be accessed by individuals who subscribe to the service by paying fees on monthly basis. The company responsible for the solution has taken a comprehensive approach in addressing all aspects of the solution's life cycle. As a result, they have full control of the entire life cycle process. This means that the company can effectively manage the solution's performance. In the design of the equipment, the issues regarding reliability, durability, and repairability are implemented. Additionally, the company provides a wide variety of services that encompass many different stages of the equipment's life cycle. These services include installing the equipment, performing both preventive and corrective maintenance, as well as making arrangements for the equipment's disposal at the end of its life. The company's comprehensive approach aims to ensure that the equipment operates efficiently throughout its entire life cycle. The company has established a reverse logistics system to address end-of-life concerns by collecting products and their components and taking responsibility for recycling them. The process is intended to reduce waste and environmental impact by ensuring proper disposal and recycling of products. The monthly subscription fees encompass all the services provided, ensuring access to clean and purified water. This means that customers do not need to worry about additional costs for water purification or any related services. By paying the subscription fees, they can ensure a supply of safe drinking water. Performing regular maintenance on the system has multiple benefits beyond just extending the equipment's lifespan. It also enhances customer loyalty and trust by ensuring that the system is functioning correctly and that any potential issues are identified and addressed promptly. Additionally, the preventive maintenance activities enable the company to establish communication with its customers, allowing them to gather data on system usage and consumption patterns. This information can be used to further improve the system and tailor it to effectively meet the needs of the customers (Jagtap 2019b). Figure 5.3 provides an illustration of the above case example.

5.2 Intersectoral Collaboration

Cross-sector partnerships between NGOs, local governments, and for-profit companies enable more effective and efficient performance of activities in the design process and life cycle stages of solutions aimed at supporting development of marginalised societies (Jagtap 2019b).

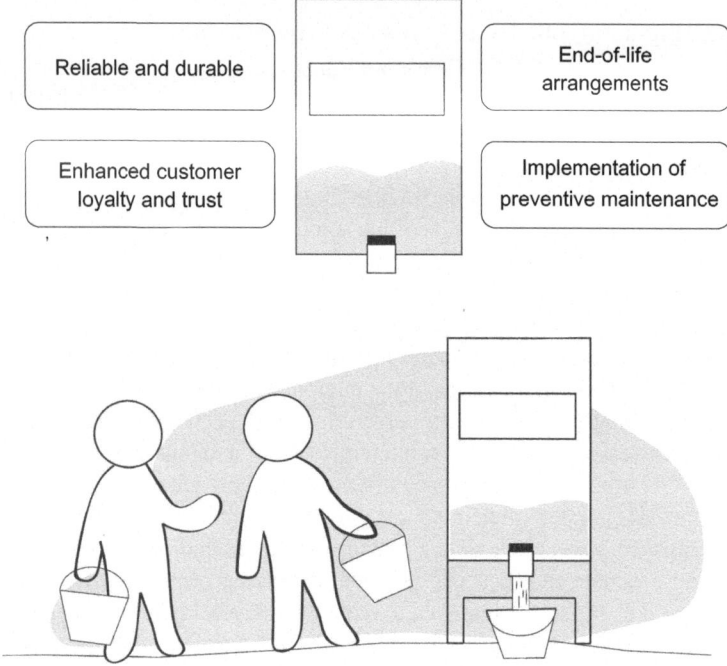

Fig. 5.3 Illustration of the 'water purification system' case example

5.2.1 Why Intersectoral Collaborative Design is Needed

In designing solutions for marginalised societies, it is crucial to involve multiple partners and utilise a variety of resources (Jagtap 2019b). This is because it is unlikely for a single partner to provide all the necessary inputs to address the numerous gaps in the value chain within these societies (Aranda-Jan et al. 2016; Jagtap 2019b). Therefore, a collaborative effort is needed to address the complex challenges faced by marginalised communities. This approach allows design, development, and implementation of a more holistic and effective solution. According to a large-scale study examining numerous design solutions across various developing nations, inclusive partnerships are fundamental to designing solutions that address the needs of marginalised societies (UNDP 2008). This study unveiled that the majority of effective solutions are designed and realised from a collaborative effort of diverse stakeholders, including, but not limited to, non-traditional partners such as governments and NGOs. Collaboration among a diverse group of stakeholders can be beneficial in combining their different skills, knowledge, and resources to overcome various obstacles and limitations present in disadvantaged communities (e.g., Hahn and Gold 2014). Collaborative efforts between businesses, NGOs, government bodies, and people living in low resource settings are crucial in addressing the needs of socioeconomically disadvantaged populations. Several studies have highlighted the importance of such partnerships in developing effective solutions (e.g., Bacchetti 2017; London and Hart 2004; Rosca et al. 2017).

Contribution of NGOs: In general, NGOs are informally and deeply ingrained in local communities (e.g., Reed and Reed 2009; Rivera-Santos and Rufín 2010). This enables them in developing relationships to better understand the needs and aspirations of marginalised populations. Their close connection to these communities makes them valuable partners in assisting other actors in gaining a comprehensive understanding of the local context (e.g., Ansari et al. 2012). By leveraging their understanding of the community and their relationships within it, NGOs can contribute to more effective design and realisation of solutions. NGOs are important in the early stages of a project because they play a crucial role of linking for-profit companies or other external organisations with the local communities (e.g., Jagtap and Larsson 2019). In addition to the above crucial contribution, NGOs can play a significant role in facilitating access to local resources, as well as in contributing to distribution and marketing efforts (e.g., Jagtap and Larsson 2019). By building on their knowledge and networks, NGOs can assist in the efficient and effective dissemination of products and services to targeted populations.

Contribution of governments: In general, governments have a key responsibility of promoting social and human development for every citizen in their country, regardless of their social standing. This also includes marginalised people, who may require additional support to ensure their inclusion in society. As such, in addition to NGOs, governments can contribute towards a broad range of activities in the process of design and implementation of solutions, including other activities associated with various life cycle phases of the solution for fulfilling needs of marginalised individuals (Palomares-Aguirre et al. 2018; Jagtap and Larsson 2019). Furthermore, governments can contribute towards dissemination of information about solutions aimed at supporting marginalised communities (Jagtap 2019b). They may also assist by providing subsidies, funds, or some relevant advice.

Contribution of businesses: In addition to local governments and NGOs, for-profit companies, based on their expertise in a broad range of fields such as technology, finance, and management, can substantially contribute towards devising solutions to support development of marginalised individuals and societies (Bengo and Arena 2013; Jagtap and Larsson 2019). They can contribute not just towards designing solutions, but also towards a broad range of life cycle activities of designed solutions. Their expertise in technology and design can be an asset in implementing new technologies. Furthermore, their network in non-local markets can also be an important asset in linking marginalised societies and non-local markets for selling locally produced products (Jagtap 2019b).

Figure 5.4 illustrates how contribution of a broad range of stakeholders contributes towards the design, development, and implementation of integrated solutions in the context of low resource settings. In this figure, NGOs are shown within the context of low resource settings as they are, in general, locally embedded. For-profit businesses are typically not embedded in the local context, and governments have the responsibility to provide basic public services to both marginalised and non-marginalised societies.

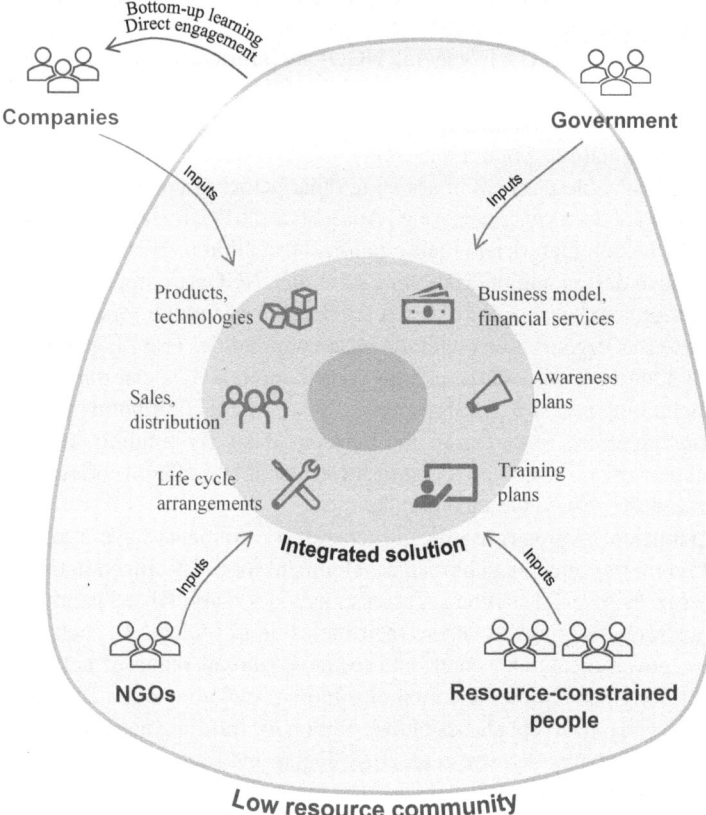

Fig. 5.4 Intersectoral collaboration and low resource settings (adapted from Jagtap and Larsson 2019)

5.2.2 How to Design Collaboratively Across Sectors

There are several methods and strategies that can support intersectoral collaborative design in the context of low resource settings (Jagtap 2019b). These are as follows:

Appropriate skills and expertise of collaborating partners: In selecting partners and developing collaborations, it is crucial to take into account knowledge, skills, and strengths possessed by each collaborating partner, and their ability to contribute towards development and implementation of solutions (e.g., Devisscher and Mont 2008; Jagtap et al. 2013).

Context-sensitive communication methods: Collaborative design in this field requires identification, creation, and implementation of context-sensitive communication methods for engaging and communicating with a broad range of partners (e.g., UNDP 2008). These communication practices ought to consider socio-cultural disparities that might exist between collaborating partners. As such, communication methods tailored to the specificities of the local context and characteristics of different partners are essential for effective collaboration in this field.

Overcoming cultural barriers: One of the crucial aspects of effectively building partnerships and organised processes in this field is to overcome cultural barriers that might exist between partners such as businesses, NGOs, and local governments (Rivera-Santos and Rufín 2010). The enablers that assist in overcoming these cultural differences are: supporting establishment of trust-based relationships, targeting the development of long-term collaboration, and developing local legitimacy.

Long-term operating horizons: In devising, implementing, and maintaining solutions for marginalised communities, it is crucial that involved partners recognise the importance and necessity of long-term operating horizons in establishing and maintaining partnerships (e.g., Jagtap et al. 2013; Jagtap 2019b). As such, stakeholders ought to engage in long-term collaboration, while recognising the positive role of patience in developing solutions for creating intended impact on the development of marginalised communities.

According to Jagtap (2021), intersectoral collaboration between partners such as NGOs, companies, and local governments serves as inputs in all phases of the design process (see Fig. 5.5). This figure also shows some methods that can support intersectoral collaboration.

Box 5.2 Intersectoral Collaboration—A Case Example

Afrique Initiatives is a company that invests in small and medium-sized enterprises in Africa (Jagtap 2019b; Jagtap and Larsson 2019). Afrique Initiatives founded Pésinet for the purpose of creating effective solutions to track the health status of children from disadvantaged families. Pésinet's goal is to improve the health outcomes of these children (since 2014, Pésinet works under the name Djantoli). Pésinet implemented the related healthcare solutions in a region called Coura. This area is located near Bamako, the capital city of Mali, and is known as a large city in Mali. By implementing their solution in this region, Pésinet was able to focus on providing better health outcomes for children in the area. The design and implementation of the solution was a collaborative effort that involved a diverse set of stakeholders. These stakeholders were the following:

- An NGO named Kafo Yeredeme Ton from Mali.
- Two universities, namely ESSEC Business School and Ecole Centrale Paris.
- Two large companies in the field of telecommunications—Alcatel-Lucent and Orange.
- A drug distributor—Medex.
- The people from the selected region Coura.
- Hospitals in the local area.

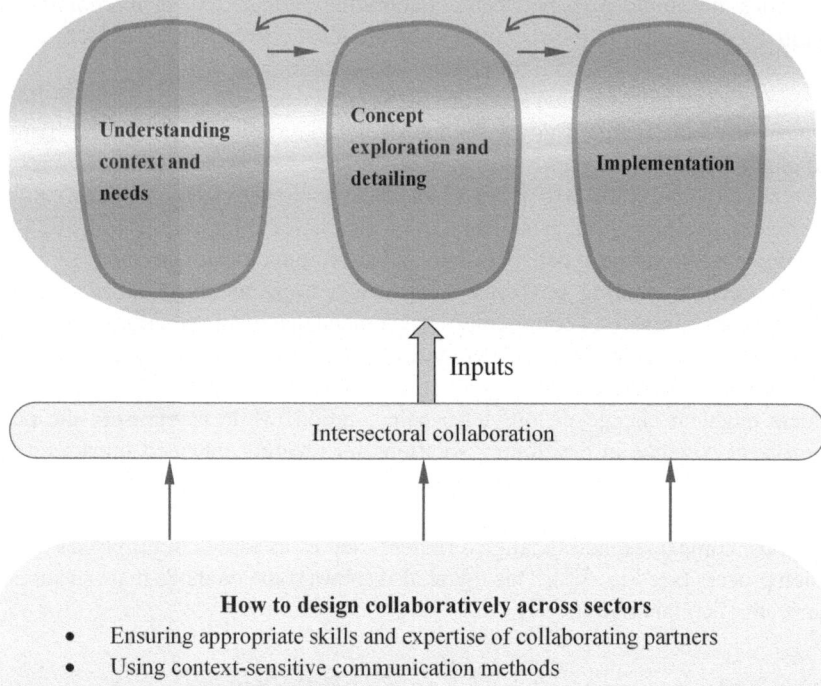

Fig. 5.5 Design process and intersectoral collaboration

These partners played a key role in the design and implementation of the solution, contributing their expertise and insights to achieve a successful outcome. A key idea in the solution is to remotely examine variation in the weight of a child. This variation in weight is used as a health-indicator. The steps in the solution are as follows:

- The child's mother subscribes to services of Pésinet by paying a nominal fee.
- A Pésinet representative visits the mother's home twice a week to weigh the child(ren) and record symptoms such as fever and diarrhoea (if any).
- The representative then transmits the weight readings and any symptoms to a local database by employing the SMS functionality of a mobile phone.
- Doctors at local healthcare facilities have access to the above database and can review the child's weight and symptoms over time. Doctor examines pattern in the weight change and other symptoms, and, if there is an anomaly,

sends an SMS to the representative mentioning that mother and child need to visit the healthcare centre. The above steps are illustrated in Fig. 5.6.

By building on their skills, resources and knowledge, Alcatel-Lucent and Orange played a crucial role in designing necessary technical systems. Economic constraints attributed to low and uncertain income of the marginalised families in the region were addressed by appropriately designed business plan as well as by the involvement of hospitals in the local area in the developed solution. The students of Ecole Centrale and ESSEC Business School contributed towards the design of the business plan. The locally embedded NGO 'Kafo Yeredeme Ton' supported other partners in gaining deeper understanding of local context. The NGO also supported design of suitable awareness programs. The successful solution was attributed to the collaborative efforts of a diverse range of stakeholders, including NGOs, for-profit businesses, and local governments (Jagtap 2019b).

5.3   Customising Project Management and Preventing Biases

Avoiding biases and adapting project management to the local specificities of marginalised societies enable effective and efficient accomplishment of design and life cycle activities (Jagtap 2019b).

5.3.1   Why to Customise Project Management and Prevent Biases

Jagtap et al.'s (2014) study revealed that there can be negative stereotypes and biases about people living in marginalised societies. The 'outsiders' can be unfamiliar about the life circumstances, needs, and ways of thinking of marginalised individuals. Their unfamiliarity, combined with their ways of thinking and working typically embedded in relatively wealthy or non-marginalised settings, can be a source of negative stereotypes about marginalised societies. For example, Jagtap et al. (2014) undertook an experimental study, using think aloud method, to compare design processes for the two distinct contexts—marginalised and non-marginalised societies. The study revealed numerous differences between these design processes. One of the key findings is that the designers engaged in designing a solution for marginalised societies did not give enough attention to needs of marginalised people regarding aesthetic and

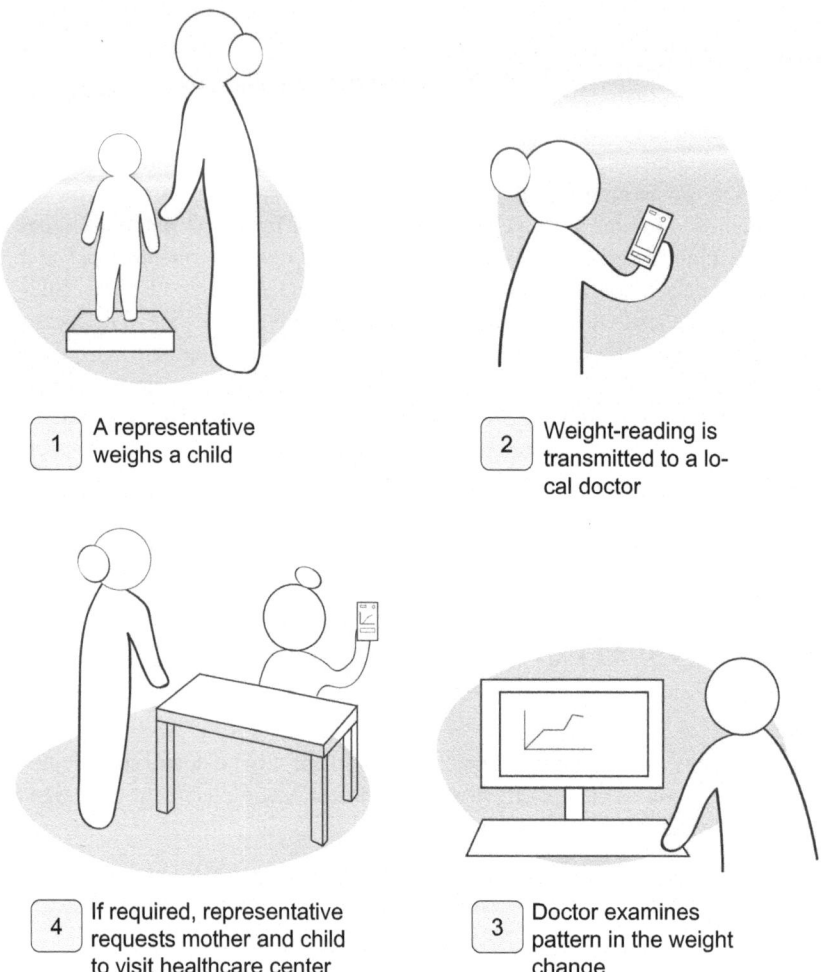

1	A representative weighs a child	2	Weight-reading is transmitted to a local doctor
4	If required, representative requests mother and child to visit healthcare center	3	Doctor examines pattern in the weight change

Fig. 5.6 Illustration of the steps in the Pésinet's case example

ergonomic qualities of products. Overlooking of these needs was despite the relevance of such needs in influencing adoption of designed solutions by marginalised people. Jagtap et al. (2015) propose that individuals from non-marginalised settings may have a tendency to think that marginalised people just have basic needs, and they may give little attention to their other needs, for example, needs about ergonomic, aesthetic, semantic, and symbolic qualities of solutions.

Previous research has suggested that people living in marginalised contexts do not just have basic survival need, and can also have other needs (e.g., Jagtap et al. 2015; Jagtap 2019b). For example, a study undertaken in Bolivia suggested that people living in marginalised settings may give priority to products showing prestige as compared to products for satisfying basic needs (Van Kempen 2009). A project in the water sector in Mexico also indicated similar results. The Water Initiative (TWI) carried out a project in order to satisfy needs of marginalised people regarding drinking water (Hart 2010). A region near Torreon in Mexico was selected for the project aimed at tackling a severe problem associated with arsenic contaminated water. In contrast to their initial assumption that the people in that area just need an affordable system to purify water, the TWI, by its engagement with the community, found that those people wished for a solution that not only purifies water, but can also make them delighted. They not just wanted purified water but also cold and good-tasting water. The above studies provide evidence that it is important to avoid biases and assumptions about needs and requirements of people living in marginalised societies in designing solutions.

While the above studies indicate biases and prejudices regarding needs and requirements of people living in marginalised settings, prejudices can also emerge from the ways in which various stakeholders, contributing to the design and development of solutions, view each other (Jagtap 2019b). Addressing various issues in marginalised settings requires holistic solutions, and designing such holistic solutions necessitates inputs from many different partners such as local governments, businesses, and NGOs (Jagtap et al. 2013). As such, the ways in which these partners view each other can also result into their biased behaviour. For example, for-profit companies may be perceived as 'venal and exploitative', governments may be considered as 'corrupt and inefficient', and NGOs may be seen as 'naïve and ineffective' (Karnani 2011). When these partners perceive each other negatively, it can be difficult for them to work together and effectively contribute to the design and implementation of a solution. To work together for contributing towards satisfying needs of marginalised individuals, they ought to positively view each other. For instance, for-profit companies can be considered as resourceful and efficient, governments as partners with the ability to formulate appropriate rules and regulations, and NGOs as partners who are active and eager to address problems.

The design projects undertaken to support development of people living in marginalised societies are inherently complex as well as ambiguous. As such, in addition to the crucial need of overcoming biases in designing solutions to support their development, it is important to adapt management of projects by taking into account specific requirements and conditions in these societies (Jagtap et al. 2014). It is important to develop necessary skills and knowledge to adapt project management to the local requirements. In order to adapt project management, the following guidelines can be useful:

- In adapting project management, explicitly stating role of each stakeholder is beneficial (Jagtap 2019b).
- The involved stakeholders ought to understand and embrace cultural differences that may exist between them as well as differences in the ways in which they communicate.
- It is crucial that these stakeholders develop skills to listen to people living in marginalised settings and employ appropriate methods to deeply understand their needs and views (e.g., Viswanathan 2016).

- Patience and taking a long-term view are important aspects because solutions developed and implemented in marginalised societies can take time to deliver results (Jagtap 2019b).

5.3.2   How to Customise Project Management and Prevent Biases

Cultural education about marginalised societies, together with respect and humility, can assist different stakeholders to avoid biases and adapt their project management (Jagtap 2019b). Furthermore, long-term planning and involvement is an essential component of project management in this field (e.g., Jagtap et al. 2014). In selecting participants for a project, it is important to consider partners willing to contribute towards developing solutions for supporting social and human development of marginalised individuals. Explicitly defining their roles and responsibilities can usefully assist in adapting project management to the local specificities (e.g., Jagtap et al. 2014; Jagtap 2019b). Additionally, involved partners need to devote efforts for gaining deeper understanding of problems in the local context. Partners, involved in the project, ought to recognise that trust- and relationship-building are time-consuming activities, and this is specifically vital in the context of marginalised settings. The bottom-up approach, articulated by Viswanathan (2016), can also usefully support in overcoming biases and preconceptions about marginalised settings, and in adapting management of project to the specificities in the target context.

According to Jagtap (2021), the tasks associated with adaptation of project management and overcoming biases serve as inputs in all phases of the design process (see Fig. 5.7). This figure also shows some methods that can support these tasks.

Box 5.3 Customising Project Management and Preventing Biases—Potential Supporting Steps

Viswanathan and Sridharan (2012) devised an interesting and useful approach known as 'bottom-up-approach' to support designing solutions for people living in marginalised settings. This approach is not just useful for stakeholders in overcoming biases and prejudices about needs of marginalised individuals, but can also support them to adapt and manage projects taking into account particular requirements and circumstances in the target context. The bottom-up approach not only allows stakeholders to gain deeper understanding of needs and problems in marginalised contexts, but also supports them in overcoming negative stereotypes about needs of marginalised individuals. The bottom-up approach consists of several steps such as:

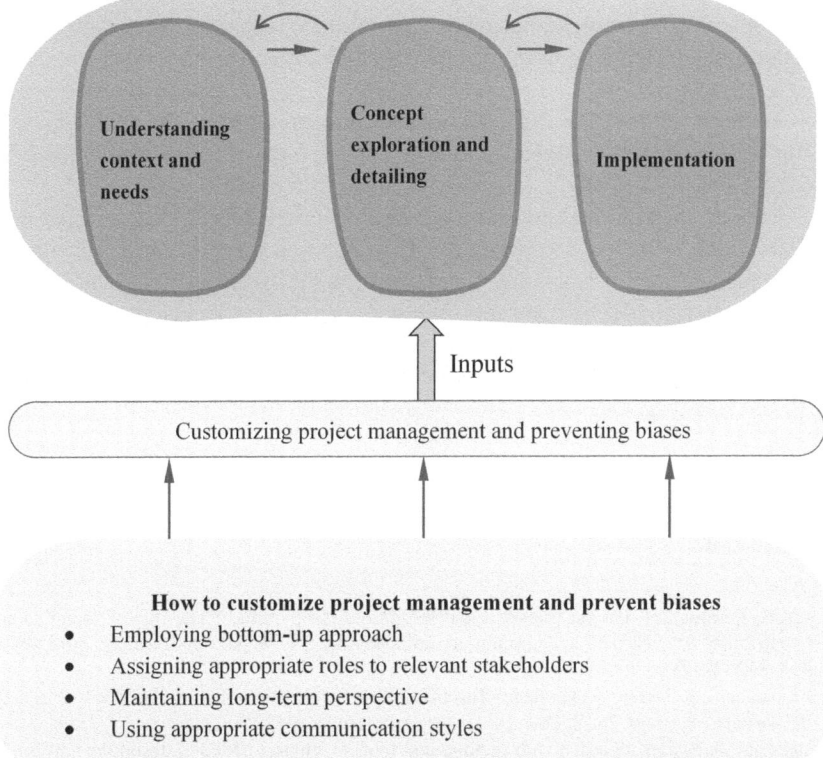

Fig. 5.7 Design process, customising project management, and preventing biases

Virtual immersion: This step helps partners in getting simulated exposure to the context of marginalised settings. This simulated exposure is assisted by the use of information about these settings, which can be in the form of text and videos.

Emersion: In this step, partners reflect on the lessons learned in the first step.

Initial idea generation and evaluation: In this step, initial ideas are generated and evaluated.

Planning for field work and immersion in the actual setting: After emersion, partners seek to gain actual, 'real-life' understanding and experience of the target context of marginalised setting. In gaining this experience, they can use multiple methods.

Reflecting on immersion: In this step, partners reflect on insights gleaned in the previous step of actual immersion and they correct their prior conceptions (if any).

Focused generation and selection of concepts: In this step, appropriate number of concepts is generated, while avoiding unmanageable generation of many ideas.

Various steps in the bottom-up approach support stakeholders not just in overcoming prejudices and biases about marginalised societies, but can also aid them in managing and adapting their projects to the particularities of the target context (Jagtap 2019b). This approach has been used in designing solutions, covering several sectors such as education, healthcare, energy, etc. Viswanathan and Sridharan (2012) have reported related examples from such sectors.

References

S. Ansari, K. Munir, T. Gregg, Impact at the 'bottom of the pyramid': The role of social capital in capability development and community empowerment. J. Manag. Stud.manag. Stud. **49**(4), 813–842 (2012)

C. Aranda-Jan, S. Jagtap, J. Moultrie, Towards a framework for holistic contextual design for low-resource settings. Int. J. Des. **10**(3), 43–63 (2016)

E. Bacchetti, A design approach with method and tools to support SMEs in designing and implementing distributed renewable energy (DRE) solutions based on sustainable product-service system (S. PSS). Procedia CIRP **64**, 229–234 (2017)

I. Bengo, M. Arena, Integrating the social dimension into new business models for energy access, in *Renewable Energy for Unleashing Sustainable Development* (Springer, Cham, 2013), pp. 203–217

T. Devisscher, O. Mont, An analysis of a product service system in Bolivia: Coffee in Yungas. Int. J. Innov. Sustain. Dev. **3**(3–4), 262–284 (2008)

S. Emili, F. Ceschin, D. Harrison, Product–service system applied to distributed renewable energy: A classification system, 15 archetypal models and a strategic design tool. Energy Sustain. Dev. **32**, 71–98 (2016)

R. Hahn, S. Gold, Resources and governance in "base of the pyramid"-partnerships: Assessing collaborations between businesses and non-business actors. J. Bus. Res. **67**(7), 1321–1333 (2014)

S. Hart, Taking the green leap to the base of the pyramid, in *Next generation Business Strategies for the Base of the Pyramid. New Approaches for Building Mutual Value* (2010), pp. 79–101

S. Jagtap, Design and poverty: A review of contexts, roles of poor people, and methods. Res. Eng. Des. **30**, 41–62 (2019a)

S. Jagtap, Key guidelines for designing integrated solutions to support development of marginalised societies. J. Clean. Prod. **219**, 148–165 (2019b)

S. Jagtap, P. Kandachar, Representing interventions from the base of the pyramid. J. Sustain. Dev. **3**(4), 58–73 (2010)

S. Jagtap, T. Larsson, Resource-limited societies, integrated design solutions, and stakeholder input. She Ji: J. Des. Econ. Innov. **5**(4), 285–303 (2019)

S. Jagtap, A. Larsson et al., Design and development of products and services at the Base of the Pyramid: A review of issues and solutions. Int. J. Sustain. Soc. **5**(3), 207–231 (2013)

S. Jagtap, A. Larsson, V. Hiort, E. Olander, A. Warell, P. Khadilkar, How design process for the base of the pyramid differs from that for the top of the pyramid. Des. Stud. **35**(5), 527–558 (2014)

S. Jagtap, A. Larsson, A. Warell, D. Santhanakrishnan, S. Jagtap, Design for the BOP and the TOP: Requirements handling behaviour of designers, in *ICoRD'15—Research into Design Across Boundaries Volume 2. Smart Innovation, Systems and Technologies*, ed. by A. Chakrabarti, vol. 35 (Springer India, 2015), pp. 191–200

S. Jagtap, Frugal-IDeM: An integrated methodology for designing frugal innovations in low-resource settings, in *Design for Tomorrow—Volume 2: Proceedings of ICoRD 2021* (Springer Singapore, 2021), pp. 41–51

K. Kamigaki, M. Matsumoto, Y.A. Fatimah, Remanufacturing and refurbishing in developed and developing countries in Asia: A case study in photocopiers. Procedia CIRP **61**, 645–650 (2017)

A. Karnani, *Fighting poverty together: Rethinking strategies for business, governments, and civil society to reduce poverty* (Palgrave Macmillan, New York, NY, 2011)

X. Lemaire, Fee-for-service companies for rural electrification with photovoltaic systems: The case of Zambia. Energy Sustain. Dev. **13**(1), 18–23 (2009)

T. London, S.L. Hart, Reinventing strategies for emerging markets: Beyond the transnational model. J. Int. Bus. Stud. **35**(5), 350–370 (2004)

D. Ness, K. Xing, A theoretical construct of a product-service systems model for low social-economic status communities. INFORMS Serv. Sci. Conf. Taipei (2010)

I. Palomares-Aguirre, M. Barnett, F. Layrisse, B.W. Husted, Built to scale? How sustainable business models can better serve the base of the pyramid. J. Clean. Prod. **172**, 4506–4513 (2018)

S.V. Ramani, S. SadreGhazi, G. Duysters, On the diffusion of toilets as bottom of the pyramid innovation: Lessons from sanitation entrepreneurs. Technol. Forecast. Soc. Chang. **79**(4), 676–687 (2012)

A. Reed, D. Reed, Partnerships for development: Four models of business involvement. J. Bus. Ethics **90**, 3–37 (2009)

Rivera-Santos, M., & Rufín, C. (2010). Global village vs. small town: Understanding networks at the Base of the Pyramid. International Business Review, 19(2), 126–139

E. Rosca, M. Arnold, J.C. Bendul, Business models for sustainable innovation—An empirical analysis of frugal products and services. J. Clean. Prod. **162**, S133–S145 (2017)

T.T. Sousa-Zomer, P.A. Cauchick-Miguel, Exploring business model innovation for sustainability: An investigation of two product-service systems. Total Qual. Manage. Bus. Excell. 1–19 (2017)

UNDP, *Creating Value for All: Strategies for Doing Business with the Poor*. United Nations Development Programme (2008)

L. Van Kempen, *Status consumption and poverty in developing countries* (VDM Publishing, Saarbrücken, 2009)

M. Viswanathan, *Bottom-up enterprise: Insights from subsistence marketplaces*. Madhu Viswanathan (2016)

M. Viswanathan, S. Sridharan, Product development for the BoP: Insights on concept and prototype development from university-based student projects in India. J. Prod. Innov. Manag.manag. **29**(1), 52–69 (2012)